SpringerBriefs in Physics

SpringerBriefs in Physics are a series of slim high-quality publications encompassing the entire spectrum of physics. Manuscripts for SpringerBriefs in Physics will be evaluated by Springer and by members of the Editorial Board. Proposals and other communication should be sent to your Publishing Editors at Springer.

Featuring compact volumes of 50 to 125 pages (approximately 20,000–45,000 words), Briefs are shorter than a conventional book but longer than a journal article. Thus Briefs serve as timely, concise tools for students, researchers, and professionals.

Typical texts for publication might include:

- A snapshot review of the current state of a hot or emerging field
- A concise introduction to core concepts that students must understand in order to make independent contributions
- An extended research report giving more details and discussion than is possible in a conventional journal article
- A manual describing underlying principles and best practices for an experimental technique
- An essay exploring new ideas within physics, related philosophical issues, or broader topics such as science and society

Briefs are characterized by fast, global electronic dissemination, straightforward publishing agreements, easy-to-use manuscript preparation and formatting guidelines, and expedited production schedules. We aim for publication 8–12 weeks after acceptance.

More information about this series at http://www.springer.com/series/8902

Tamás Sándor Biró · Antal Jakovác

Emergence of Temperature in Examples and Related Nuisances in Field Theory

 Springer

Tamás Sándor Biró
H.A.S. Wigner Research Centre
for Physics
Budapest, Hungary

Antal Jakovác
Institute of Physics
Roland Eötvös University
Budapest, Hungary

ISSN 2191-5423   ISSN 2191-5431   (electronic)
SpringerBriefs in Physics
ISBN 978-3-030-11688-0   ISBN 978-3-030-11689-7   (eBook)
https://doi.org/10.1007/978-3-030-11689-7

Library of Congress Control Number: 2019931531

This Springer imprint is published by the registered company Springer Nature Switzerland AG
The registered company address is: Gewerbestrasse 11, 6330 Cham, Switzerland

Preface

Field theory is often surrounded with a mystical aura. People who are not acquainted with it often think that it can solve deep problems which, without field theory, could not even have been addressed. A few others who have been using field theory methods for decades sometimes allude to a *fin de siècle* feeling of not being there further challenges for field theory. Of course, both these sentiments are misleading: As a theory of physics, and relying as it does on the concept of continuous space and time while at the same time being confronted with the quantum physical nature of observable quantities, field theory still has some fundamental challenges before it.

One of these challenges is *to understand the emergence of complexity on the fundamental level*. Instances of complexity in the behavior of interacting elementary fields constitute problems which call for explanation on the fundamental level. These are among other nontrivial phase structures of elementary matter at high energy density or an atypical emergence of statistical properties, e.g., when an apparent temperature is proportional to a constant acceleration.

The classical theory of heat and temperature, i.e., thermodynamics, has already obtained a deeper level description, first in the kinetic theory by Maxwell and Boltzmann, then in its statistical generalization by Gibbs, Planck, and Einstein, and its further connection to information theory by Maxwell's demon, Szilard, and Shannon. In the recent past, there has even been some speculation about 'explaining' Einstein's equations, which describe the reaction of space-time structures to the presence of matter, from the concept of entropy assigned to event horizons. A key step toward this description is the analogy between acceleration and temperature, already stated by William Unruh in 1975. However, its appearance seemed to be entangled with some deep philosophical issues: More specifically, a constantly accelerating observer would see a 'heated vacuum' and hence detect the presence of an energy *characteristic of a finite temperature, emerging from nothing*. The question of where this temperature comes from has not been answered to this today.

A finite temperature taken as a fact, established by some external agents and kept to a constant value by some miracle, can be built into the methods of field theory. There have been several decades of work on this, accumulating a certain level of knowledge, and also, since the 1980s and 1990s, several good textbooks devoted to finite temperature field theory [1–6] and its numerical simulation by lattice field theory [7–10]. Its nonperturbative, numerical study, on the other hand, has grown into a scientific industry with the fast pace of emerging computational and information technology over the past 50 years. We feel that there is no urgent need to increase the number of textbooks on this.

However, responding to the invitation of Balasubramanian Ananthanarayan, Editor for the Springer Brief series of concise physics textbooks, we focus here on a slightly provocative approach to field theory methods: not simply asking how they should be used if a finite temperature is given, but rather trying to understand how the phenomenon of temperature emerges physically for elementary fields. What is the kinetic field theory? How should we handle the fluctuation of energy and other noise? How does this harmonize (or not?) with standard field theory methods? Therefore, we shall endeavor here to include already existing but not necessarily published knowledge on this fundamental problem, from our own experience as theoretical physicists working on related questions. This has led to our suggestion to write about the 'emergence of temperature and related nuisances' in particle physics and field theory.

The structure of this brief is as follows. We begin by comparing energy variances of thermal and quantum origins, discussing the Unruh temperature which emerges due to a constant acceleration of the observer. Then, the lessons for the quantum energy variance will be generalized to the treatment of off-mass-shell effects via a density operator, a Wigner function, and eventually with the help of propagators. Nonequilibrium propagators and the Keldysh two-time formalism of quantum field theory are then introduced, and in a follow-up chapter, their application to linear response theory, describing near-equilibrium behavior. Finally, we discuss situations where the prerequisites for using the exponential Boltzmann factor are not satisfied, whence a more general treatment of noise is called for. To round off our cruise through these troubled waters, we present somewhat more detailed analyses of slow-mode dynamics in Yang-Mills plasmas, the statistics of local energy packets in classical lattice gauge theory, and a model Lagrangian with analytically given spectral function and the resulting thermodynamical equation of state.

We would like to thank the support of the Hungarian National Innovation Office supervised by the Ministry of Innovation and Technology under the project NKFIH/K123815.

Budapest, Hungary   Tamás Sándor Biró
November 2018   Antal Jakovác

References

1. J.I. Kapusta, *Finite-Temperature Field Theory*. Cambridge Monographs on Mathematical Physics (Cambridge University Press, Cambridge, 1993)
2. J.I. Kapusta, C. Gale, *Finite-Temperature Field Theory: Principles and Applications* (Cambridge University Press, Cambridge, 2006)
3. A. Wipf, *Statistical Approach to Quantum Field Theory in the Framework of Thermo Field Dynamics*. Springer Lecture Notes in Physics, vol. 864 (Springer, Berlin, 2012)
4. A. Das, *Finite Temperature Field Theory* (World Scientific, Singapore, 1997)
5. M. Kaku, *Quantum Field Theory (A Modern Introduction)* (Oxford University Press, Oxford, 1993)
6. S. Mallik, S. Sarkar, *Hadrons at Finite Temperature* (Cambridge University Press, Cambridge, 2016)
7. M. Creutz, *Quarks, Gluons and Lattices* (Cambridge University Press, Cambridge, 1983)
8. C. Gattringer, C.B. Lang, *Quantum Chromodynamics on the Lattice*. Springer Lecture Notes in Physics, vol. 788 (Springer, Berlin, 2010)
9. H.J. Rothe, *Lattice Gauge Theories (An Introduction)*. Lecture Notes in Physics, vol. 74 (World Scientific, Singapore, 2005)
10. T.J. Hollowood, *Renormalization Group and Fixed Points in Quantum Field Theory*. Springer Briefs in Physics (Springer, Berlin, 2013)

Contents

Chapter 1
Quantum Uncertainty and Unruh Temperature

Temperature as a measure of fluctuations is well known among physicists [1, 2]. More specifically, its value – measured in Boltzmann's constant units—is connected to the variance of the energy (after subtraction of the collective motion), $\Delta E \sim k_B T$. Usually, only the statistical variance is considered, this being due to internal atomic motion, according to the kinetic theory of heat [3, 4]. In particle physics the analog phenomenon is the motion of elementary particles in an ensemble, as in the kind of hot fireball constructed in high energy accelerator experiments [5]. In field theory, if enough energy is present, the number of particles cannot be fixed, and this quantity also fluctuates in a way that depends on further circumstances. A repeated collision event—even at the same accelerator energy—will produce different numbers of hadrons each time. Here the energy per particle fluctuates due to the fluctuating number of particles and also due to the fluctuating individual kinetic energies of those particles [6, 7].

If one attempts to describe a quantum system using classical concepts, all measurements deliver fluctuating values for the classical quantity, too. This is not usually called a 'temperature'. Indeed, this phenomenon is traditionally referred to as 'uncertainty'. The famous uncertainty relation is an inequality providing a lower limit for the product of variances of two such quantities observed simultaneously. In order to identify the similarities between these two seemingly different types of fluctuations, we have to raise some fundamental questions about the relationship between classical and quantum mechanics. It is important to be aware of the key steps which led to the use of operator algebra in quantum calculations and to the 'shut up and calculate' attitude [8].

In this chapter we begin with a concise review of the beginnings of quantum mechanics. Schrödinger's variational principle, which relies on but also partially abandons the classical Hamilton–Jacobi equation of motion, led to his quantum equation and to the concept of a complex amplitude wave function for calculating probabilities. As we shall see below, this approach really abandons the exact fulfillment of classical equations of motion, and eventually also abandons the idea that point-like entities can be taken as the basis for correct physics. The exact equation of motion, for

© The Author(s), under exclusive licence to Springer Nature Switzerland AG 2019

T. S. Biró and A. Jakovác, *Emergence of Temperature in Examples and Related Nuisances in Field Theory*, SpringerBriefs in Physics,
https://doi.org/10.1007/978-3-030-11689-7_1

integrable cases, is equivalent to an exact relation between the Hamiltonian function of generalized coordinates and momenta and the energy. The Copenhagen school, on the other hand, identified the energy with the Hamiltonian, and therefore—still satisfying the Schrödinger equation rather than the Hamilton–Jacobi equation—the only mathematical way out is to use operators instead of functions, and in particular Hermitian operators instead of real-valued functions, in the description of physical reality on the atomic level and below.

Finally, the uncertainty relation emerges as a consequence of the algebra of non-commuting Hermitian operators, and not only among the basic phase space variables of generalized coordinates and momenta, but between all quantities described by Hermitian operators [9–13]. In this chapter, we explore in particular the case of constant force fields, when the commutator of the momentum operator and the Hamiltonian is proportional to the force, which is a non-zero constant. Therefore the energy and the momentum both have a variance of quantum origin, bounded from below by this constant; or equivalently by the constant acceleration of a massive particle in such a force field. Interestingly enough, the energy variance at a given momentum variance of just creating a particle of mass m, viz., $\Delta P \sim mc$, will then be proportional to this acceleration. This fact opens a new window for considering a temperature, proportional to a constant acceleration, and through this for viewing quantum uncertainty as a particular 'temperature effect'.

1.1 Variational Principle Behind Quantum Mechanics and Operators

It is simplest to start with Schrödinger's derivation of his famous equation from a variational principle. This can be found in the original publications [14–17]. Here we present a simplified and, for the purpose of the present discussion, somewhat modified version of the original chain of thought.

Let us consider first the stationary case: this shows the background concept in the cleanest way. Classical mechanics is described by using an action integral and a Hamiltonian function of the basic phase space variables, viz., generalized coordinates, Q, and generalized momenta, P. For stationary cases the energy, E, is constant and therefore it is enough to use the reduced action, $S_{\text{red}} = S + tE$. The conjugate momentum is obtained as

$$P = \frac{\partial S}{\partial Q}, \tag{1.1}$$

and the Hamilton–Jacobi equation is simply

$$H\left(Q, \frac{\partial S}{\partial Q}\right) - E = 0. \tag{1.2}$$

The value of the Hamilton function is equal to the energy, which is a constant of motion. In the world of quantum physics this classical equation is not fulfilled. However, we do not abandon it entirely, we just look for a minimal violation of it in the integral sense. The quantity measuring this violation is defined as

$$\mathcal{K}_{\text{stac}} \equiv \int dQ\, w(Q) \left[H\!\left(Q, \frac{\partial S}{\partial Q}\right) - E \right] , \tag{1.3}$$

and we are looking for a theory which minimizes this expression. The integration weight factor, $w(Q)$, remains unknown for the moment. If it is positive semi-definite and can be normalized to

$$\int dQ\, w(Q) = 1 , \tag{1.4}$$

we may later interpret $w(Q)$ as a probability density of a given local violation of the classical mechanical equations, but this interpretation is not obligatory.

How do we obtain Schrödinger's equation from this? The key is to view the action in the so-called eikonal picture,

$$S = k \ln \psi , \tag{1.5}$$

with k a constant to be determined later. Aiming at a description of the wave nature of particle propagation in quantum mechanics, we allow for complex values in this formula: the eikonal, the action, and even the coefficient can be complex. Now we rewrite (1.3) for a particular class of simple Hamiltonians:

$$H(Q, P) = \frac{1}{2m} |P|^2 + V(Q) . \tag{1.6}$$

The quantum violation measure becomes

$$\mathcal{K}_{\text{stac}} = \int dQ\, w(Q) \left[\frac{|k|^2}{2m} \left| \frac{1}{\psi} \frac{\partial \psi}{\partial Q} \right|^2 + V(Q) - E \right] . \tag{1.7}$$

Now, with the purpose in mind of obtaining an equation *linear* in the complex eikonal, $\psi(Q)$, for the variational minimum, we choose

$$w(Q) \equiv |\psi|^2 = \psi^* \psi . \tag{1.8}$$

Indeed, in this case,

$$\mathcal{K}_{\text{stac}}^{\text{lin}} = \int dQ \left[\frac{|k|^2}{2m} \left| \frac{\partial \psi}{\partial Q} \right|^2 + [V(Q) - E]|\psi|^2 \right] \tag{1.9}$$

is a quadratic functional of $\psi(Q)$, and its variation delivers exactly the Schrödinger equation:

$$\frac{\delta \mathcal{K}_{\text{stac}}^{\text{lin}}}{\delta \psi^*} = -\frac{|k|^2}{2m} \frac{\partial^2 \psi}{\partial Q^2} + \left[V(Q) - E \right] \psi = 0 \, . \tag{1.10}$$

If we wish to interpret this equation in terms of classical concepts, a formal analogy to (1.2) and (1.6) suggests writing it in the form

$$\left[\frac{1}{2m} \hat{P}^2 + V(\hat{Q}) - E \right] \psi = 0 \, . \tag{1.11}$$

Comparing (1.11) with (1.10) reveals that the 'hatted' momentum is a differential operator,

$$\hat{P} = \frac{|k|}{\mathrm{i}} \frac{\partial}{\partial Q} \, , \tag{1.12}$$

while $\hat{Q} = Q$ can be used. As an immediate consequence, since differential operators can be represented by matrices in a given function basis, the order of $\hat{P}$ and $\hat{Q}$ matters. Their commutator, viz.,

$$[\hat{P}, \hat{Q}] \equiv \hat{P}\hat{Q} - \hat{Q}\hat{P} = \frac{|k|}{\mathrm{i}} \, , \tag{1.13}$$

is considered today as a fundamental relation of quantum physics. Indeed, comparison with experimental results suggests that $|k| = \hbar$. However, this does not fix the phase of the complex coefficient, k, only its magnitude.

   To find out more about the phase, we repeat the above argument for the time-dependent case. The classical equation, viz.,

$$H \left(\frac{\partial S}{\partial Q}, Q \right) + \frac{\partial S}{\partial t} = 0 \, , \tag{1.14}$$

will not be satisfied, but a variational principle is formulated which extremizes its integral with respect to the generalized coordinate, Q, and with respect to time t :

$$\mathcal{K} = \int \mathrm{d}t \int \mathrm{d}Q \left[H \left(\frac{\partial S}{\partial Q}, Q \right) + \frac{\partial S}{\partial t} \right] |\psi|^2 \tag{1.15}$$

should have a vanishing first functional derivative with respect to the function $\psi(Q, t)$ obtained from (1.5):

$$\psi = \mathrm{e}^{S/k} \, . \tag{1.16}$$

For the simple Hamiltonian class discussed above, the violation measure of classical mechanics becomes

$$\mathcal{K} = \int \mathrm{d}t \int \mathrm{d}Q \left[\frac{|k|^2}{2m} \frac{\partial \psi}{\partial Q} \frac{\partial \psi^*}{\partial Q} + V(Q) \psi^* \psi + k \psi^* \frac{\partial \psi}{\partial t} \right] \, . \tag{1.17}$$

Now we have to regard $\mathcal{H}$ as a bilinear functional of the complex function pair $\psi(Q, t)$ and $\psi^*(Q, t)$, and require that the respective functional derivatives are mutually consistent.[1] Hence,

$$\frac{\delta \mathcal{H}}{\delta \psi^*} = -\frac{|k|^2}{2m}\frac{\partial^2 \psi}{\partial Q^2} + V(Q)\psi + k\frac{\partial \psi}{\partial t} = 0 \tag{1.18}$$

must agree with the complex conjugate of

$$\frac{\delta \mathcal{H}}{\delta \psi} = -\frac{|k|^2}{2m}\frac{\partial^2 \psi^*}{\partial Q^2} + V(Q)\psi^* - k\frac{\partial \psi^*}{\partial t} = 0 . \tag{1.19}$$

This can only work if $k = -k^*$, so k must be pure imaginary. Following convention, one chooses $k = \hbar/\mathrm{i}$, resulting in

$$\hat{P} = \frac{\hbar}{\mathrm{i}}\frac{\partial}{\partial Q} , \qquad \psi = \mathrm{e}^{\mathrm{i}S/\hbar} . \tag{1.20}$$

1.2 Quantum Uncertainty Algebra

Once quantum physics is modelled by a mathematical structure using operators, the next step is to realize that 'physical' quantities, which can be tested in measurements, are described by a special class of operators, namely the Hermitian operators, since these correspond to real-valued functions in classical physics. These operators, also called self-adjoint operators, have only real eigenvalues when acting on eigenstates [18].

Many fundamental and interesting questions are raised by this special model of the nature of quantum phenomena. The states on which these operators act span a Hilbert vector space, with mathematically well-defined properties [19]. These are discussed in numerous textbooks and research articles. Here we focus on the (in)famous uncertainty relation. For this purpose let us investigate two Hermitian operators, representing two different physical, measurable quantities, denoted by $\hat{A}$ and $\hat{B}$. Their respective expectation values can be set to zero without loss of generality and the conclusions below are then easily extended from $\hat{A} \equiv \hat{a} - \langle a \rangle$ to the general case.

As a general operator we first consider $\hat{C} = \hat{A} + \mathrm{i}\hat{B}$, with $\hat{A}$ and $\hat{B}$ Hermitian. The product of any operator with its adjoint must have a non-negative expectation value:

$$\left\langle \left(\hat{A} + \mathrm{i}\hat{B}\right)\left(\hat{A}^\dagger - \mathrm{i}\hat{B}^\dagger\right)\right\rangle \geq 0 . \tag{1.21}$$

Expanding and using the Hermitian property, i.e., $\hat{A}^\dagger = \hat{A}$ and $\hat{B}^\dagger = \hat{B}$, this reads

---

[1]This is the functional analogue of the Cauchy–Riemann relation known for complex analytic functions.

$$\langle \hat{A}^2 + i(\hat{B}\hat{A} - \hat{A}\hat{B}) + \hat{B}^2 \rangle \geq 0 . \tag{1.22}$$

Collecting the obviously positive expectation values on the left-hand side, we arrive at

$$\langle \hat{A}^2 \rangle + \langle \hat{B}^2 \rangle \geq \langle i[\hat{A}, \hat{B}] \rangle . \tag{1.23}$$

The same derivation is valid if we interchange the roles of A and B. Thus, from

$$\langle (\hat{B} + i\hat{A})(\hat{B}^\dagger - i\hat{A}^\dagger) \rangle \geq 0 , \tag{1.24}$$

it follows that

$$\langle \hat{B}^2 \rangle + \langle \hat{A}^2 \rangle \geq \langle i[\hat{B}, \hat{A}] \rangle . \tag{1.25}$$

These results can be collected into a single formula:

$$\langle \hat{A}^2 \rangle + \langle \hat{B}^2 \rangle \geq \left| \langle i[\hat{A}, \hat{B}] \rangle \right| . \tag{1.26}$$

Note that, since $\hat{A}$ and $\hat{B}$ are both Hermitian, all quantities in this inequality are now non-negative reals, so this is a nontrivial consequence of the operator algebra.

Recalling that the above derivation was made for $\hat{A} = \hat{a} - \langle \hat{a} \rangle$ and $\hat{B} = \hat{b} - \langle \hat{b} \rangle$, also Hermitian operators, we realize that the squared expectation values are the *variances* of the original quantities, while the commutator term is the same. Recalling also the well-known inequality between the arithmetic and geometric means, the minimum of the left-hand side is given by

$$\min \left(\Delta a^2 + \Delta b^2 \right) = 2\Delta a \cdot \Delta b . \tag{1.27}$$

The final meaning of (1.26) is then that the product of variances has to satisfy the following inequality:

$$\Delta a \cdot \Delta b \geq \left| \left\langle \frac{i}{2}[\hat{a}, \hat{b}] \right\rangle \right| . \tag{1.28}$$

Applying this basic result to the momentum and position, or more generally to a canonically conjugate pair of Hermitian operators, $\hat{P}$ and $\hat{Q}$, we obtain the quantum uncertainty relation due to Heisenberg:

$$\Delta P \cdot \Delta Q \geq \frac{\hbar}{2} . \tag{1.29}$$

But there is much more in this. Considering the class of additive Hamiltonians,

$$\hat{H} = T(\hat{P}) + V(\hat{Q}) , \tag{1.30}$$

we arrive at the following uncertainty relation between energy and momentum measurements:

$$\Delta E \cdot \Delta P \geq \left| \left\langle \frac{i}{2} [\hat{P}, \hat{H}] \right\rangle \right| = \frac{\hbar}{2} \left| \langle \hat{F} \rangle \right| , \tag{1.31}$$

with the usual definition of force in conservative potentials:

$$F \equiv -\frac{\partial V}{\partial Q} . \tag{1.32}$$

In particular for a constant force field, like weak gravity, the energy and momentum 'uncertainty' (variance) is bounded from below by a constant. Naturally, in the absence of the constant force, this bound is zero, and it is perfectly possible to have a 'sharp' dispersion relation, $E(P)$.

In order to investigate this correspondence more deeply, let us choose a photon in a weak gravitational field as an example. The gravitational redshift reduces the energy to

$$E = \hbar\omega \sqrt{1 - \frac{2GM}{c^2 r}} \approx \hbar\omega \left(1 - \frac{GM}{c^2 r} \right) , \tag{1.33}$$

for a photon with frequency ω moving radially at distance r from a mass M, according to the Schwarzschild metric. Regarding the correction to the free photon energy, $\hbar\omega$, as a Newtonian gravitational potential, it is easy to derive a gravitating 'mass' for the photon:

$$m = \frac{\hbar\omega}{c^2} . \tag{1.34}$$

The weak gravitational force acting on the photon is given by

$$F = -\frac{\partial V}{\partial r} \approx \frac{GM}{c^2 r^2} \hbar\omega = mg . \tag{1.35}$$

Here the constant acceleration due to gravity $g = GM/r^2$ is used as an abbreviation. According to (1.31), we obtain

$$\Delta E \cdot \Delta P \geq \frac{\hbar}{2} \frac{g}{c^2} \hbar\omega . \tag{1.36}$$

Noting that the expectation value of the photon energy in this approximation is $\langle E \rangle = \hbar\omega$, and that the photon's velocity in vacuum is c, whence $\Delta E = c\Delta P$, we conclude finally that there is a curious lower bound:

$$\frac{\Delta E^2}{\langle E \rangle} \geq \frac{\hbar g}{2c} . \tag{1.37}$$

This lower bound is, believe it or not, proportional to the Unruh temperature[2] [20–23]:

$$T_{\text{Unruh}} = \frac{\hbar g}{2\pi c} .$$ (1.38)

Just out of curiosity, let us compare this lower bound for the quantum uncertainty variance with a thermal one. For an order of magnitude estimate, we may use the Boltzmann approximation, and for more precise calculations, the Bose distribution for photons with zero rest mass. Quite generally, in D spatial dimensions, we have

$$\left\langle E^K \right\rangle = \frac{1}{Z} \int_0^\infty E^K \, e^{-E/T} \, E^{D-1} \mathrm{d}E = T^K \frac{\Gamma(D+K)}{\Gamma(D)} .$$ (1.39)

Using this, we have $\left\langle E \right\rangle = DT$ and $\left\langle E^2 \right\rangle = D(D+1)T^2$, and therefore $\Delta E^2 = DT^2$. From this it follows that

$$\frac{\Delta E^2}{\left\langle E \right\rangle} = T .$$ (1.40)

If one wishes to view the temperature precisely as a measure of the energy variance in a Boltzmann distribution of energies, then, disregarding the factor π, the Unruh temperature is nothing other than 'the temperature' associated with the quantum state with minimal uncertainty.

1.3   Quantum Fluctuations: A Generalized Temperature

In this section we test how general the identification of a temperature due to the variance measure of quantum fluctuations can be, as introduced in the particular example above. *More precisely, we establish a connection between the deviation of the energy from its classical value and the variance calculated in the operator picture.* In the relativistic generalization of this phenomenon, the energy deviation (off-energy-shell behavior) will be replaced by the deviation from the classical dispersion relation between energy and momentum, which eventually occurs as an 'off-mass-shell' feature.

A particularly transparent approach to this property of quantum physics can be given by a discussion of the Schrödinger principle, introduced in Sect. 1.1, in terms of 'Madelung–Bohm' variables. The complex wave function in this picture is factorized into a magnitude and a complex phase term:

$$\psi = R \, e^{i\alpha/\hbar} .$$ (1.41)

---

[2]This will be derived in Sect. 1.4. Here we use units such that $k_B = 1$.

Correspondingly, the complex action, the generalization of the classical real action, splits into

$$S = \alpha + \frac{\hbar}{i} \ln R \ . \tag{1.42}$$

In this notation, the gradient of the complex action is given by

$$\nabla S = \nabla\alpha + \frac{\hbar}{i} \frac{\nabla R}{R} \ , \tag{1.43}$$

and the variational action behind Schrödinger's equation takes the form

$$\mathscr{K} = \int \left[\frac{|\nabla\alpha|^2}{2m} + V(x) + \frac{\partial\alpha}{\partial t} \right] R^2 \ d^3x \ dt + \frac{\hbar^2}{2m} \int |\nabla R|^2 \ d^3x \ dt. \tag{1.44}$$

The imaginary term stemming from $\partial S/\partial t$ contains $R\partial R/\partial t$, which is a total time derivative. It contributes to the integral only at the time endpoints and has been neglected here.

This variational quantum action can now be varied with respect to the functions α and R. The first variation delivers a continuity equation:

$$\frac{\delta\mathscr{K}}{\delta\alpha} = -\nabla\left(\frac{\nabla\alpha}{m} R^2\right) - \frac{\partial}{\partial t} R^2 = 0 \ . \tag{1.45}$$

It is intriguing in this result that it formally resembles a flow of $\rho = R^2$, a positive definite real density with the classical velocity $\mathbf{v} = \langle \hat{\mathbf{P}} \rangle/m = \nabla\alpha/m$:

$$\frac{\partial\rho}{\partial t} + \nabla(\rho\mathbf{v}) = 0 \ . \tag{1.46}$$

This result also helps to interpret $R^2 = |\psi|^2$. Considering the spatial integral of the continuity equation (1.45), viz.,

$$\frac{d}{dt}\left(\int R^2 \ d^3x\right) + \int \nabla\left(R^2 \frac{\nabla\alpha}{m}\right) d^3x = 0 \ , \tag{1.47}$$

the second integral is solely a surface term and can be set to zero. As a consequence, the integral

$$\int R^2 \ d^3x = \int |\psi|^2 \ d^3x \tag{1.48}$$

is a constant in time. The traditional choice is the value 1 for this constant, and therefore $\rho = R^2$ is interpreted as a probability density function. Schrödinger, who had originally designed his equation to describe the dynamics of a single electron, called this the 'probability of the electron being there'. Strictly speaking, only its integrals in a finite spatial domain can be probabilities.

The other variation leads to a form that is no less interesting:

$$\frac{|\nabla \alpha|^2}{2m} + V(x) + \frac{\partial \alpha}{\partial t} = \frac{\hbar^2}{2m} \frac{\nabla^2 R}{R} . \tag{1.49}$$

In the stationary case $\frac{\partial \alpha}{\partial t} = -E$, the left-hand side of the above expression is exactly the deviation of the energy from its classical value, expressed in terms of the real action α. Formally, for $\hbar = 0$, it is the classical equation of motion and α is the classical action.

Let us now compare these results with the variances calculated in the operator picture. In particular, we are interested in the expectation value and variance of the momentum operator $\hat{P}$:

$$\langle \hat{\mathbf{P}} \rangle = \int \psi^* \frac{\hbar}{i} \nabla \psi \, \mathrm{d}^3 x = \int R^2 \nabla \alpha \, \mathrm{d}^3 x , \tag{1.50}$$

up to a surface term, and

$$\langle \hat{\mathbf{P}}^2 \rangle = \int R^2 (\nabla \alpha)^2 \, \mathrm{d}^3 x + \hbar^2 \int (\nabla R)^2 \, \mathrm{d}^3 x . \tag{1.51}$$

Here the first integral represents the classical momentum squared integral weighted by R^2, while the second term is formally proportional to $\hbar^2$ and is a genuine quantum contribution. This second term can also be estimated by using (1.49) after partial integration

$$\int (\nabla R)^2 \, \mathrm{d}^3 x = - \int R \nabla^2 R \, \mathrm{d}^3 x + \text{surface term} . \tag{1.52}$$

With the energetic 'off-shell'-ness taken into account, we arrive at

$$\langle \hat{\mathbf{P}}^2 \rangle = \int R^2 (\nabla \alpha)^2 \, \mathrm{d}^3 x - 2m \int R^2 \left[\frac{(\nabla \alpha)^2}{2m} + V(x) - E \right] \mathrm{d}^3 x . \tag{1.53}$$

Although it is now hidden in the formula, the off-shell contribution is proportional to $\hbar^2$.

Concentrating now on the variance of the momentum operator, we have two contributions:

$$\Delta P^2 = \Delta P_{\text{class}}^2 + \Delta P_{\text{off-shell}}^2 , \tag{1.54}$$

with

$$\Delta P_{\text{class}}^2 = \int R^2 (\nabla \alpha)^2 \, \mathrm{d}^3 x - \left[\int R^2 \nabla \alpha \, \mathrm{d}^3 x \right]^2 \tag{1.55}$$

and

$$\Delta P_{\text{off-shell}}^2 = - \int R^2 \left[(\nabla \alpha)^2 + 2m(V(x) - E) \right] \mathrm{d}^3 x . \tag{1.56}$$

In order to deepen our understanding of this result, let us consider two examples. First the case $R =$ constant, $E = (\nabla\alpha)^2/2m + V$ and $\Delta P^2 = 0$, characterizes a momentum eigenstate. In this case the continuity equation delivers $\nabla^2\alpha = 0$, whose solution is a linear dependence of the phase of the wave function on the position vector: $\alpha = \hbar\mathbf{k}\cdot\mathbf{r} + \varphi_0$. Since $E = |\mathbf{k}|^2/2m + V$ is also constant, this is possible only in constant potentials, i.e., without any force acting on the quantum particle. Then due to $E = -\partial\alpha/\partial t$, there is a spatially constant but time dependent contribution to the quantum phase: $\alpha = \hbar\mathbf{k}\cdot\mathbf{r} - Et$. Using Planck's law, $E = \hbar\omega$, we obtain

$$\psi = \text{const }e^{i(\mathbf{k}\cdot\mathbf{r}-\omega t)} . \tag{1.57}$$

In our next example, we consider quantum binding in a co-moving frame. In this case $\langle \hat{\mathbf{P}} \rangle = \langle \nabla\alpha \rangle = 0$. Now R must be time-independent due to the flow continuity equation, but there may be a space dependence. The 'off-shell' Eq. (1.49) now reduces to

$$V(\mathbf{r}) - E = \frac{\hbar^2}{2m}\frac{\nabla^2 R}{R} . \tag{1.58}$$

Let us restrict this discussion to a Coulomb potential, $V = -e^2/r$, and consider the ground state of a hydrogen atom (one-electron problem in a given static potential). In this case the off-shell equation delivers

$$\left(R'' + \frac{2}{r}R'\right) + \frac{2m}{\hbar^2}\left(E + \frac{e^2}{r}\right)R = 0 . \tag{1.59}$$

As is well known, the ground state solution is sought in the exponential form: $R(r) = K\,e^{-r/a}$. In this case $R' = -R/a$ and $R'' = R/a^2$, and one concludes that

$$\left[\frac{1}{a^2} - \frac{2}{ar} + \frac{2m}{\hbar^2}\left(E + \frac{e^2}{r}\right)\right]R = 0 . \tag{1.60}$$

Obviously, this is only possible if

$$E = -\frac{\hbar^2}{2ma^2} \quad \text{and} \quad e^2 = \frac{\hbar^2}{ma} , \tag{1.61}$$

retrieving Bohr's results [24]. The Bohr radius is given as $a = \hbar^2/me^2$. Let us now check the variance of the momentum operator, whose expectation value was already set to zero. We obtain

$$\Delta P^2 = \langle \hat{\mathbf{P}}^2 \rangle = \hbar^2\int |\nabla R|^2\,d^3x = \frac{\hbar^2}{a^2} . \tag{1.62}$$

The conclusion, $\Delta P = \hbar/a$ while $\Delta E = 0$, confirms that the expectation values of the force vector components are zero in such a spherically symmetric state.

Indeed, any Hermitian operator constructed as a function of phase space variables, $\hat{B} = \mathscr{B}(\hat{P}, \hat{Q})$, evolves in time in a way determined by the Hamiltonian:

$$\frac{\hbar}{i} \frac{d}{dt} \hat{B} = \left[\hat{H}, \hat{B} \right] . \tag{1.63}$$

According to the general result on the uncertainty, we have

$$\Delta E \cdot \Delta B \geq \frac{\hbar}{2} \left| \left\langle \frac{d}{dt} \hat{B} \right\rangle \right| . \tag{1.64}$$

This form allows us to extract a time uncertainty associated with the Hermitian operator (physical quantity) $\hat{B}$:

$$\Delta t_B \equiv \frac{\Delta B}{\left| \left\langle d\hat{B}/dt \right\rangle \right|} , \tag{1.65}$$

and reformulate the energy uncertainty in the form:

$$\Delta E \cdot \Delta t_B \geq \frac{\hbar}{2} . \tag{1.66}$$

This lower limit is valid for any quantity described by $\mathscr{B}(\hat{P}, \hat{Q})$. In particular the time uncertainty associated with the momentum component is its spread ΔP divided by the absolute value of the expectation value of the corresponding force component, while the time uncertainty associated with the position vector component is its spread ΔQ divided by the velocity component, and so on.

At this point, however, we encounter a further difficulty. Velocity vector components depend on the observer's own velocity, and force components depend on the observer's acceleration. Apparently, we can always find a reference frame where the velocity vanishes, whence $\Delta t_Q = \infty$ and ΔE becomes compatible with zero; and similarly, we can always find an accelerating frame equivalent to an external gravitational field in general relativity where the force vanishes (a freely falling system), whence $\Delta t_P = \infty$ and $\Delta E = 0$ is possible. Indeed, the commutator expression underlying the uncertainty relation must reflect a value invariant with respect to the observer's motion. For the energy and momentum commutator it is 'automatically' given if they are regarded as components of the same four-momentum vector. Their Lorentz transformation,

$$E' = \gamma (E - vP) , \qquad P' = \gamma \left(P - \frac{v}{c^2} E \right) , \tag{1.67}$$

delivers

$$[E', P'] = \left[\gamma (E - vP), \gamma (P - vE/c^2) \right] = \gamma^2 [E, P] + \gamma^2 \frac{v^2}{c^2} [P, E] . \tag{1.68}$$

Then, introducing the Lorentz factor for γ with the property $\gamma^2(1 - v^2/c^2) = 1$, this delivers

$$[E', P'] = [E, P] . \tag{1.69}$$

The case with the original Heisenberg uncertainty is more delicate. For a Lorentz transformation of the coordinate operator, $\hat{x}$, one has to conjecture the existence of a time-coordinate operator, $\hat{t}$, because of the linear mixing for an inertial observer:

$$x' = \gamma(x - vt) , \qquad t' = \gamma(t - vx/c^2) . \tag{1.70}$$

According to this, the commutators transform as follows:

$$[P', x'] = \gamma^2[P, x] - \gamma^2 v[P, t] - \gamma^2 \frac{v}{c^2}[E, x] + \gamma^2 \frac{v^2}{c^2}[E, t] ,$$

$$[E', t'] = \gamma^2[E, t] - \gamma^2 v[P, t] - \gamma^2 \frac{v}{c^2}[E, x] + \gamma^2 \frac{v^2}{c^2}[P, x] . \tag{1.71}$$

This transformation conserves the difference of commutators,

$$[E', t'] - [P', x'] = [E, t] - [P, x] , \tag{1.72}$$

but does not leave the individual commutators invariant. Requiring, on the other hand, that the canonical commutator in both systems be set at $[P', x'] = [P, x] = \hbar/i$, we conclude that we must have $[E', t'] = [E, t]$. Putting these relations back into (1.71), we obtain

$$v([P, x] + [E, t]) = c^2[P, t] + [E, x] . \tag{1.73}$$

If quantum uncertainty is a Lorentz invariant phenomenon, then the above equality should hold for arbitrary values of the observer's velocity v. This is only possible if both factors are zero in the original frame:

$$[P, x] + [E, t] = 0 ,$$
$$c^2[P, t] + [E, x] = 0 . \tag{1.74}$$

From the first equation, we immediately obtain $[E, t] = i\hbar$ and hence $\Delta E \cdot \Delta t \geq \hbar/2$. However, the second equality also has to be satisfied. Interpreting the time operator as multiplication by the time coordinate, t, the energy operator becomes $\hat{E} = i\hbar \partial/\partial t$, and both $[P, t]$ and $[E, x]$ vanish. Therefore the energy operator cannot be a Hamiltonian with P-dependent kinetic energy term in this case. Such paradoxes around the time and energy operators render non-relativistic quantum mechanics inadequate for describing physics when high velocities are involved.

Since the original quantum mechanics does not respect the precepts of special and general relativity, we have to enlarge our perspective here and turn to a relativistic version of quantum mechanics. For the time being, we consider the special relativistic extension of Schrödinger's variational principle. This leads to a complex scalar field and its relativistic action.

The Lagrangian density for a free complex scalar field (no interaction, no force) is given by

$$\mathscr{L} = \partial_i \psi^* \partial^i \psi - \left(\frac{mc}{\hbar}\right)^2 \psi^* \psi \,. \tag{1.75}$$

The variational principle extremizes the action $\mathscr{S} = \int \mathscr{L} d^4 x$, with the spacetime differential $dx^i = (cdt, d\mathbf{r})$ and the invariant 4-volume form $d^4 x = cdt\, d^3 r$. We again use the Madelung (polar) representation of the complex scalar as follows:

$$\psi = \frac{\hbar}{\sqrt{mc}} R\, e^{i\alpha/\hbar} \,. \tag{1.76}$$

This form has been chosen so that the mass term in the Lagrangian looks as close as possible to the classical description of a relativistic point with mass m. In particular, with this choice one has

$$\left(\frac{mc}{\hbar}\right)^2 \psi^* \psi = mcR^2 \,. \tag{1.77}$$

Its integral, the 'static' part of the free action only, becomes

$$-mc \int R^2 d^4 x = -\int dt \left(mc^2 \int R^2 d^3 x\right) = -\int mc^2 d\tau \tag{1.78}$$

if, in the co-moving frame, one ensures

$$\int R^2 \, d^3 x = 1 \,. \tag{1.79}$$

This is the connection with the normalized wave function interpretation.

We now consider the derivative terms. Any directional derivative in spacetime is given by

$$\partial_i \psi = \left(\frac{1}{R}\partial_i R + \frac{i}{\hbar}\partial_i \alpha\right) \psi \,. \tag{1.80}$$

It is convenient to take the classical four-momentum to be the four-gradient of the real action term α, and we shall set $P_i = \partial_i \alpha$ from now on. Then the Lagrangian becomes

$$\mathscr{L} = \frac{\hbar^2}{mc}\partial_i R \cdot \partial^i R + \frac{R^2}{mc}\left[P_i \cdot P^i - (mc)^2\right] \,. \tag{1.81}$$

We see the same structure as for the non-relativistic Schrödinger action: there is a term without explicit reference to $\hbar$ corresponding to the classical equation of motion (which is not zero now) multiplied by the wave function density, R^2, and there is an explicitly quantum-looking addition containing derivatives of R.

This action has a $U(1)$ symmetry due to the indeterminacy of the absolute quantum phase, and this corresponds to the following Noether four-current:

$$J^k = \frac{i}{2\hbar} \left(\psi \partial^k \psi^* - \psi^* \partial^k \psi \right) = R^2 \frac{P^k}{mc} . \tag{1.82}$$

It is proportional to the classical four-velocity $u^i = P^i/(mc)$. The variation of the action $\mathscr{S}$ with the Lagrange density (1.75) delivers its conservation law:

$$\frac{\delta \mathscr{S}}{\delta \alpha} = -\partial_k J^k = 0 . \tag{1.83}$$

The other field equation, obtained by variation with respect to R, describes the off-mass-shell measure of a freely propagating mass m quantum field:

$$\frac{\delta \mathscr{S}}{\delta R} = \frac{2R}{mc} \left[P_i P^i - (mc)^2 \right] - \frac{2\hbar^2}{mc} \partial_i \partial^i R = 0 . \tag{1.84}$$

The off-shellness is thus given by

$$P_i P^i - (mc)^2 = \hbar^2 \frac{\Box R}{R} . \tag{1.85}$$

1.4   Unruh Temperature

It seems that the energy uncertainty is bound to the special relativistic treatment of quantum phenomena. On the other hand, the characteristic magnitude of ΔE for photons, as extreme relativistic 'particles', revealed a temperature proportional to the acceleration in a constant external force field. In this section we review a derivation of the Unruh temperature.

We consider a pointlike, monochromatic light source moving with constant acceleration, g, on a straight line with relativistic speed. As a function of the proper time τ, its trajectory (a so-called Rindler trajectory) is given by[3]

$$t = \frac{c}{g} \sinh \frac{g\tau}{c} , \qquad x = \frac{c^2}{g} \left(\cosh \frac{g\tau}{c} - 1 \right) . \tag{1.86}$$

The retarded phase from such a source amounts to

$$\varphi = \omega_0(t - x/c) = \frac{\omega_0 c}{g} \left(1 - e^{-g\tau/c} \right) . \tag{1.87}$$

---

[3] Note that g/c is a frequency and c/g has dimensions of time.

Since this phase changes during the motion as τ advances, a distant observer detects a complete spectrum of frequencies due to a continuously changing Doppler effect. The complex amplitude of a Fourier analysis is proportional to

$$A(\omega) \propto \int_{-\infty}^{+\infty} e^{i\varphi(\tau)} e^{i\omega\tau} d\tau . \tag{1.88}$$

The observed spectral intensity will be proportional to $|A|^2$. The integral (1.88) can be carried out formally by introducing the variable $z = e^{-g\tau/c}$, which proves to be the relativistic Doppler redshift factor. In terms of this, the complex spectral amplitude (neglecting the pure phase factor $e^{i\omega_0 c/g}$) becomes

$$A(\omega) \propto \frac{c}{g} \int_0^{\infty} z^{-i\omega c/g-1} e^{-i\omega_0 zc/g} dz. \tag{1.89}$$

This is nothing other than Euler's Gamma integral:

$$A(\omega) \propto \frac{c}{g} \left(-i\frac{\omega_0 c}{g} \right)^{-i\omega c/g} \Gamma(-i\omega c/g) . \tag{1.90}$$

Here we use the fact that $(-i)^{-i} = (e^{-i\pi/2})^{-i} = e^{-\pi/2}$. The intensity is proportional to

$$|A(\omega)|^2 \propto \frac{c^2}{g^2} \Gamma(-i\omega c/g)\Gamma(i\omega c/g) e^{-\pi\omega c/g} . \tag{1.91}$$

The product of conjugate Gamma functions, viz.,

$$\Gamma(ia)\Gamma(-ia) = \frac{\pi}{a \sinh(a\pi)} ,$$

can be inserted to obtain

$$|A(\omega)|^2 \propto \frac{2\pi c}{g\omega} \frac{1}{e^{2\pi\omega c/g} - 1} . \tag{1.92}$$

Since the number density of photons is counted by utilizing $\omega^2|A(\omega)|^2 = \omega n(\omega)$, one concludes that this spectrum resembles a Planck radiation spectrum for $n(\omega)$ at a temperature, T, satisfying

$$k_B T = \frac{\hbar}{2\pi c} g . \tag{1.93}$$

Finally, we note that, continuing the Rindler trajectory to imaginary proper time, $\tau = i\hbar\beta = i\hbar/k_B T$, it is periodic for the above Unruh temperature, since $g\tau/c = 2i\pi$.

Summarizing this chapter, a variance in measured energy values can have three sources:

- classical thermal fluctuations,
- quantum uncertainty, and
- constant acceleration in an external field.

The second and third mechanisms lead in some cases to observations which are indistinguishable from a thermal spectrum and hence from a telemetric temperature measurement. Changing the external force, however, the spectrum also changes and one will no longer observe a sharp temperature value. Nevertheless, the constant relativistic acceleration of a monochromatic source and the quantum uncertainty of energy in a constant force field both act in the same direction: towards a generalization of the notion of temperature measured by the energy variance.

References

1. H.B. Callen, *Thermodynamics and Introduction to Thermostatistics* (Wiley, New York, 1985)
2. H.J.D. Miller, J. Anders, Energy-temperature uncertainty relation in quantum thermodynamics. Nat. Commun. **9**, 2203 (2018)
3. B.H. Lavenda, *Statistical Physics: A Probabilistic Approach* (Wiley Interscience, New York, 1991)
4. R.L. Liboff, *Kinetic Theory: Classical, Quantum and Relativistic Descriptions* (Springer, Berlin, 2006)
5. W. Florkowski, *Phenomenology of Ultra-Relativistic Heavy-Ion Collisions* (World Scientific, Singapore, 2010)
6. L. Stodolsky, Temperature fluctuations in multiparticle production. Phys. Rev. Lett. **75**, 1044 (1995)
7. T.S. Biro, Z. Schram, L. Jenkovszky, Entropy production during hadronization of a quark-gluon plasma. EPJ A **54**, 17 (2018)
8. N.D. Mermin, Could Feynman have said this? Phys. Today **57**, 10 (2004)
9. W. Heisenberg, The physical content of quantum kinematics and mechanics. Z. Phys. **43**, 172 (1927)
10. M. Osawa, Universally valid reformulation of the Heisenberg uncertainty principle on noise and disturbance measurements. Phys. Rev. A **67**, 042105 (2003)
11. J.S. Briggs, A derivation of the time-energy uncertainty relation. J. Phys. Conf. Ser. **99**, 012002 (2008)
12. P. Busch, P.P. Lahti, R. Werner, Proof of Heisenberg's error-disturbance relation. Phys. Rev. Lett. **111**, 160405 (2013)
13. D.M. Appleby, Quantum errors and disturbances: response to Busch, Lahti and Werner. Entropy **18**, 174 (2016)
14. Erwin Schrödinger, An undulatory theory of the mechanics of atoms and molecules. Phys. Rev. **28**, 1049 (1926)
15. Erwin Schrödinger, *Collected Papers on Wave Mechanics* (AMS Chelsea Publishing, New York, 1982)
16. E. Schrödinger, Quantisierung als Eigenwertproblem (Erste Mitteilung). Ann. Phys. **79**, 361 (1926)
17. E. Schrödinger, Quantisierung als Eigenwertproblem (Zweite Mitteilung). Ann. Phys. **79**, 488 (1926)

18. F. Zang, *Matrix Theory: Basic Results and Techniques* (Springer, Berlin, 2011)
19. N. Young, *An Introduction to Hilbert Space* (Cambridge University Press, Cambridge, 1988)
20. W.G. Unruh, *Notes on black hole evaporation*, Phys. Rev. D **14**, 870 (1976)
21. W.G. Unruh, R.M. Wald, What happens when an accelerating observer detects a Rindler particle. Phys. Rev. D **29**, 1047 (1984)
22. L.C.B. Crispino, A. Higuchi, G.E.A. Matsas, The Unruh effect and its applications. Rev. Mod. Phys. **80**, 787 (2008)
23. T.S. Biro, *Is There a Temperature? Fundamental Theories of Physics 1014* (Springer, Berlin, 2011)
24. Niels Bohr, On the constitution of atoms and molecules. Philos. Mag. **26**, 1 (1913)

Chapter 2
Off-Shell Transport Dynamics

In Chap. 1 we discussed examples of quantum uncertainty, and also the energy variance and its resemblance to a temperature when special relativity is taken into account. We have demonstrated that the complex scalar free field theory, when viewed in terms of amplitude and phase variables, exhibits a coupling between the off-mass-shell relation of the classical four-momentum and the quantum scale variation of the magnitude of the quantum probability density. The classical dispersion relation $P_i P^i - (mc)^2 = 0$ is no longer valid, unless one deals with plane waves of constant amplitude in space and time.

This phenomenon is related to the energy variance, which simulates a temperature effect, comparable in scale to the Unruh temperature. In this chapter we shall try to identify how general this phenomenon is in quantum field theory, and see how the off-shellness enters the solution of dynamical problems when energy and momentum are physically transported. We start with a short description of another technique used to treat quantum systems, which exploits the Wigner function and the related Wigner and Weyl transformations [1–4]. In a given approximation, energy and momenta close to the classical values are transported in space and time, while their quantum uncertainty (which can also be treated as if it were a statistical variance) will be transported in space and time by other, coupled equations. Finally, the relation between the quantum uncertainty and a width in the spectral function will be analyzed in a certain approximation in order to find out how off-shell-ness and a related generalized temperature describing the relative energy variance, $\Delta E^2 / \langle E \rangle$, can be read off from the spectral properties of quantum fields.

2.1 Density Matrix and Wigner Transform

A quantum system is seldom in an eigenstate of a simple Hamiltonian. It is usually in a superposition of such eigenstates, and the complex coefficients of the individual contributions change in time as the system evolves:

T. S. Biró and A. Jakovác, *Emergence of Temperature in Examples and Related Nuisances in Field Theory*, SpringerBriefs in Physics,
https://doi.org/10.1007/978-3-030-11689-7_2

$$|\psi\rangle = \sum_m C_m(t) |m\rangle \ . \tag{2.1}$$

The density operator for such a pure state is simply a projector:

$$\hat{\rho} \equiv |\psi\rangle \langle\psi| \ . \tag{2.2}$$

The probability of finding a state $|\psi\rangle$ in the state $|m\rangle$ is given by $|\langle\psi|\,m\rangle|^2 = |C_m|^2$, but the density operator is expressed by a double sum

$$\hat{\rho} = \sum_{n,m} C_m^* C_n |n\rangle \langle m| \ . \tag{2.3}$$

Here all non-diagonal terms $n \neq m$ occur in general with complex coefficients. This operator is by construction Hermitian, i.e., $\hat{\rho}^\dagger = \hat{\rho}$. In the so-called *random phase approximation*, the non-diagonal elements average out to zero, and one is left with

$$\hat{\rho} \equiv \sum_m |C_m|^2 |m\rangle \langle m| \ , \tag{2.4}$$

where $|C_m|^2$ can be interpreted as the probability of being in state m.[1] The density operator is used to calculate the expectation value of any other operator in a general state. Based on this definition, it is easy to see that

$$\langle \hat{A} \rangle = \mathrm{tr}(\hat{\rho}\hat{A}) = \sum_{m,n} \langle n|\,\hat{\rho}\,|m\rangle \langle m|\,\hat{A}\,|n\rangle \ . \tag{2.5}$$

When $\hat{A}$ is replaced by the identity operator, it follows that $\mathrm{tr}(\hat{\rho}) = \sum_n |C_n|^2 = 1$.

In particular, in a coordinate eigenstate basis, a diagonal density operator can be expressed as a matrix with elements

$$\rho(x_1, x_2) = \langle x_1|\,\hat{\rho}\,|x_2\rangle = \sum_m |C_m|^2 \langle x_1|\,m\rangle \langle m|\,x_2\rangle \ . \tag{2.6}$$

If viewed in a basis of momentum eigenstates with momenta $\hbar k_m$, this resembles a Fourier expansion:

$$\rho(x_1, x_2) = \sum_m |C_m|^2 \, e^{ik_m(x_2 - x_1)} \ . \tag{2.7}$$

The probability of having a momentum $\hbar k_m$, on the other hand, is given by an inverse Fourier transform. More precisely, we define the Wigner function

---

[1]It is obvious that the density operator goes beyond the notion of probability, involving quantum interference, too. More general forms than (2.2) describe 'mixed states' or equivalently a statistical ensemble of quantum states.

$$W(x, p) = \int e^{\frac{i}{\hbar} p \cdot q} \rho \left(x + \frac{q}{2}, x - \frac{q}{2} \right) dq \ . \tag{2.8}$$

In this definition $x = (x_1 + x_2)/2$ and $q = x_1 - x_2$. The above structure of the Wigner transformation is the same in any dimensions, and in particular in 4-dimensional spacetime viewed in terms of four-momentum eigenstates. The inverse of this integral transformation, the Weyl transform, interprets the coordinate basis density matrix as if the Wigner function were the probability of having momentum p in the relative motion:

$$\rho(x_1, x_2) = \int W \left(\frac{x_1 + x_2}{2}, p \right) e^{\frac{i}{\hbar} p \cdot (x_1 - x_2)} \frac{dp}{2\pi\hbar} \ . \tag{2.9}$$

However, the Wigner function is not always positive, so it cannot be a probability. Nevertheless, it is possibly the concept that comes closest to the classical phase space density. Its marginal integrals are probability densities of the respective wave function representations:

$$\int W(x, p) \frac{dp}{2\pi\hbar} = |\varphi(x)|^2 \ , \quad \int W(x, p) dx = |\tilde{\varphi}(p)|^2 \ . \tag{2.10}$$

The above definition of the Wigner function is based on the coordinate representation of the wave function. It is therefore necessary to generalize (2.8) and (2.9) to arbitrary representations of quantum states. Put another way, a Wigner operator is needed. It is not difficult to see that the construction of the Wigner function is related to the shift operator frequently used in quantum optics when acting on coherent states [5]: the definition involves equal shifts of the coordinate in opposite directions.
    The shift operator is given by

$$\hat{D}(p, q) \equiv e^{\frac{i}{\hbar} \left(p \hat{Q} - q \hat{P} \right)} \ . \tag{2.11}$$

A general quantum state $|\psi\rangle$ can be expanded in terms of coordinate eigenstates, the coefficients constitute the traditional wave function:

$$|\psi\rangle = \int dx \, \varphi(x) \, |x\rangle \ . \tag{2.12}$$

In order to find the general form of the Wigner operator, we first study the action of the shift operator on the coordinate eigenstate:

$$\hat{D}(p, q) |x\rangle = e^{\frac{i}{\hbar} p \left(x - \frac{q}{2} \right)} |x - q\rangle \ . \tag{2.13}$$

The transition matrix element is thus given by

$$\langle x_1 | \hat{D}(p, q) | x_2 \rangle = e^{\frac{i}{\hbar} p \left(x_2 - \frac{q}{2} \right)} \delta(x_1 - x_2 + q) \ . \tag{2.14}$$

The expectation value of the above shift operator in the general quantum state therefore reads

$$\langle\psi|\,\hat{D}(p,q)\,|\psi\rangle = \int dx_1 \int dx_2 \, \varphi^*(x_1)\varphi(x_2) \, e^{\frac{i}{\hbar}p\left(x_2-\frac{q}{2}\right)} \, \delta(x_1 - x_2 + q) \,. \quad (2.15)$$

Using the variables $x = (x_1 + x_2)/2$ and $\Delta = x_1 - x_2$, one of the integrations can be carried out and one gets

$$\langle\psi|\,\hat{D}(p,q)\,|\psi\rangle = \int dx \, e^{\frac{i}{\hbar}px} \, \varphi^*\left(x - \frac{q}{2}\right) \varphi\left(x + \frac{q}{2}\right) \,. \quad (2.16)$$

This result is not actually the Wigner function, because it represents a Fourier transformation in the common coordinate and not in the relative one. However, this analogue quantity is convenient when describing the formation of bound states between two particles.

The Wigner function can be obtained if one performs a parity change in the $\langle\psi|$ state. Then, since $\langle x|\,\hat{\Pi} = \langle -x|$, the integral over the sum of coordinates can be eliminated and the Fourier transformation over the difference remains:

$$\langle\psi|\,\hat{\Pi}\hat{D}(p,q)\,|\psi\rangle = \int dx_1 \int dx_2 \, \varphi^*(x_1)\varphi(x_2) \, e^{\frac{i}{\hbar}p\left(x_2-\frac{q}{2}\right)} \, \delta(-x_1 - x_2 + q) \,. \quad (2.17)$$

The final result with half sum and difference coordinates is now a Wigner function

$$\langle\psi|\,\hat{\Pi}\hat{D}(p,q)\,|\psi\rangle = \frac{1}{2}\int d\Delta \, e^{-\frac{i}{2\hbar}p\Delta} \, \varphi^*\left(\frac{\Delta}{2}+\frac{q}{2}\right)\varphi\left(\frac{q}{2}-\frac{\Delta}{2}\right) = \frac{1}{2}W\left(\frac{p}{2},\frac{q}{2}\right) \,. \quad (2.18)$$

We have thus identified the Wigner operator in terms of the parity change and the coherent state shift operator as

$$\hat{W}(p,q) \equiv 2\hat{\Pi}\,\hat{D}(2p, 2q) \,. \quad (2.19)$$

In a general coherent state of the kind often treated in quantum optics, the Wigner function is defined by the expectation value of the shift operation [6]:

$$W(x,p) = 2\text{tr}\left(\hat{\rho}\hat{\Pi}\,\hat{D}(2\alpha)\right) \,, \quad (2.20)$$

using the shift operator

$$\hat{D}(z) = e^{z\hat{a}^\dagger - z^*\hat{a}} \,, \quad (2.21)$$

with $z = 2\alpha$, $\alpha = (x + ip)/\sqrt{2}$, and $[\hat{a}, \hat{a}^\dagger] = 1$. Note that the shift operator transforms the Fock vacuum state into a Glauber coherent state

$$|z\rangle = \hat{D}(z)\,|0\rangle = e^{-|z|^2/2} \sum_n \frac{z^n}{\sqrt{n!}}\,|n\rangle \,. \quad (2.22)$$

The density operator is reconstructed from the Wigner function as

$$\hat{\rho} = \int \frac{dp\,dx}{\pi\hbar} W(x, p)\hat{D}\left((x + ip)\sqrt{2}\right)\hat{\Pi} . \tag{2.23}$$

Field theory can be built using a basis of coherent states, as well as with an orthogonal system of plane wave states. A field operator in the Heisenberg picture, viz.,

$$\hat{\Phi}(x) \equiv \sum_k \sqrt{\frac{\hbar c}{2\omega_k}}\left[h_k^*(x)a_k^\dagger + h_k(x)a_k\right] , \tag{2.24}$$

is defined with help of a mode sum over states labelled by k, creation and annihilation operators of quanta in each of the modes, $a_k^\dagger$ and a_k, respectively, and a set of spacetime functions, $h_k(x)$, which form a complete basis:

$$\sum_k h_k(x)h_k^*(y) = \delta(x - y) . \tag{2.25}$$

For a plane wave basis, we also have orthogonality, i.e.,

$$\int dx\, h_k^*(x)h_\ell(x) = \delta_{k\ell} , \tag{2.26}$$

but it is not necessary. The corresponding canonically conjugate field operator is given by

$$\hat{\Pi}(x) \equiv \sum_k \sqrt{\frac{\hbar\omega_k}{2c}}i\left[h_k^*(x)a_k^\dagger - h_k(x)a_k\right] . \tag{2.27}$$

This definition delivers the basic commutator relation

$$\left[\hat{\Pi}(x), \hat{\Phi}(y)\right] = \frac{\hbar}{i}\delta(x - y) . \tag{2.28}$$

The expectation values of the basic field operators in a Glauber coherent state, for which $a_k|\psi\rangle = z_k|\psi\rangle$, are real-valued classical functions:

$$\langle\psi|\,\hat{\Phi}(x)\,|\psi\rangle = \sum_k \sqrt{\frac{\hbar c}{2\omega_k}}\left[h_k^*(x)z_k^* + h_k(x)z_k\right] = \varphi(x) ,$$

$$\langle\psi|\,\hat{\Pi}(x)\,|\psi\rangle = \sum_k \sqrt{\frac{\hbar\omega_k}{2c}}i\left[h_k^*(x)z_k^* - h_k(x)z_k\right] = \pi(x) . \tag{2.29}$$

However, the correlations between the field operators at two different points already contain quantum contributions. Here we just give the simplest examples:

$$\langle\psi|\,\hat{\Phi}(x)\hat{\Phi}(y)\,|\psi\rangle = \varphi(x)\varphi(y) + \sum_k \frac{\hbar c}{2\omega_k} h_k(x)h_k^*(y)\,,$$

$$\langle\psi|\,\hat{\Pi}(x)\hat{\Pi}(y)\,|\psi\rangle = \pi(x)\pi(y) + \sum_k \frac{\hbar\omega_k}{2c} h_k(x)h_k^*(y)\,. \qquad (2.30)$$

These quantities are in general complex-valued, and not symmetric under exchange of the space points x and y. In particular, the commutator of the same field operator at different points is non-vanishing in such a state:

$$\langle\psi|\,\left[\hat{\Phi}(x),\hat{\Phi}(y)\right]\,|\psi\rangle = \sum_k \frac{\hbar c}{2\omega_k}\left[h_k(x)h_k^*(y) - h_k(y)h_k^*(x)\right]\,, \qquad (2.31)$$

and purely imaginary.[2] This latter property is true, not only for coherent states, but also for arbitrary ones; it will provide the basis for defining the spectral function in the next section.

Applying this to a plane wave basis in one dimension and replacing the mode sum over k by an integral, we find, for highly relativistic particles and utilizing $\omega_k = |k|$,

$$\Delta\Phi(x) \cdot \Delta\Phi(y) \geq \frac{\hbar}{2}\sum_k \frac{\sin k(x-y)}{\omega_k}\,. \qquad (2.32)$$

The right-hand side of (2.31) and inequality (2.32) is actually the spectral function, which we shall discuss further below. The fact that the product of variances has an $(x-y)$-dependent lower bound implies correlations between field values at separate spacetime points. However, this may violate micro-causality: there must be no propagation of information for spacelike separated points x and y. In particular, at the same time, $t_x = t_y$, the above quantity is zero whenever only proper Lorentz transformations are allowed, i.e., Lorentz transformations without reflections. Then $S(x-y) = S(y-x)$, and therefore $S = 0$ for spacelike separations.

2.2 Dispersion Relation Entangles with Transport

A typical field theory equation, derived by varying the classical action, has the following structure:

$$\Box_1\Phi_1 + \frac{(mc)^2}{\hbar^2}\Phi_1 = J_1 + \int_3 \Sigma_{13}\Phi_3\,. \qquad (2.33)$$

Here the indices provide a compressed notation for the dependence on the corresponding spacetime coordinates, $x_i = (ct_i, \mathbf{x}_i)$, and $\Box_i$ is the d'Alembert operator

---

[2]It reflects the essentially delocalized nature of coherent states.

(relativistic wave operator), constructed from the second derivatives with respect to the spacetime coordinates:

$$\Box_i \equiv \frac{1}{c^2}\frac{\partial^2}{\partial t_i^2} - \frac{\partial^2}{\partial x_i^2} - \frac{\partial^2}{\partial y_i^2} - \frac{\partial^2}{\partial z_i^2} \, . \tag{2.34}$$

J is a general source term, 'probing' the response of the field Φ. Finally, $\Sigma_{ij} = \Sigma(x_i, x_j)$ is a so-called self-energy term, describing the interaction of the Φ-field with itself. We proceed with a general form of $\Sigma(x_i, x_j)$, although homogeneity is very often assumed. The indexed integral sign denotes integration over the corresponding four-volume in spacetime.

Instead of solving for $\Phi(x)$, one usually solves for the Green function, defined as the 'operator' transforming the external source, $J(x)$, to the response solution, $\Phi(x)$:

$$\Phi_1 = \int_2 G_{12} J_2 \, . \tag{2.35}$$

Putting this back into (2.33) and using the properties of the Dirac delta,

$$\left[\Box_1 + \left(\frac{mc}{\hbar}\right)^2 \right] \int_2 G_{12} J_2 = \int_2 \delta_{12} J_2 + \int_{2,3} \Sigma_{13} G_{32} J_2 \, , \tag{2.36}$$

one derives the corresponding equation for the Green function:

$$\left[\Box_1 + \left(\frac{mc}{\hbar}\right)^2 \right] G_{12} = \delta_{12} + \int_3 \Sigma_{13} G_{32} \, . \tag{2.37}$$

Now swapping the spacetime coordinates x_1 and x_2, the symmetry obtained by combining time reversal, parity, and charge conjugation can be exploited, setting $G_{21} = G_{12}^*$. In this way, one obtains

$$\left[\Box_2^* + \left(\frac{mc}{\hbar}\right)^2 \right] G_{12} = \delta_{12} + \left(\int_3 \Sigma_{23} G_{31} \right)^* , \tag{2.38}$$

where the asterisk denotes complex conjugation. Subtracting (2.38) from (2.37), one arrives at the Kadanoff–Baym equation [7–9], which does not contain the Dirac-delta contribution (characteristic of wave scattering problems):

$$\left(\Box_1 - \Box_2^* \right) G_{12} = \int_3 \Sigma_{13} G_{32} - \left(\int_3 \Sigma_{23} G_{31} \right)^* . \tag{2.39}$$

In the following, we derive from the original dynamical equation for the Green function (2.37). This is an evolution equation in spacetime which contains contributions reminiscent of the Vlasov equation [10, 11] (used to describe transport phenomena in plasmas) and will show off-shell-type quantum effects. To obtain this result, we take the Wigner transform of the Green function (exact retarded propagator) and the self-energy (interaction) term.

We use the four-dimensional spacetime Wigner transforms of the Green functions and self-energies:

$$W(x, p) = \int dq \; e^{ipq/\hbar} \, G(x + q/2, x - q/2) \,. \tag{2.40}$$

We denote the Wigner transform of the self-energy by $U(x, p)$. The Weyl transform reverses the Wigner transform:

$$G(x_1, x_2) = \int \frac{dp}{2\pi\hbar} \, e^{-\frac{ip}{\hbar}(x_1 - x_2)} \, W\left(\frac{x_1 + x_2}{2}, p\right) \,. \tag{2.41}$$

Using this formula one can easily derive the action of the spacetime derivative on the Green function. It will be equivalent to a complex operation on the Wigner function:

$$\partial_1 G_{12} \longrightarrow \left(-\frac{ip}{\hbar} + \frac{1}{2}\partial_x\right) W\left(\frac{x_1 + x_2}{2}, p\right) \,. \tag{2.42}$$

Using this information it is straightforward to derive the following Wigner transform of the left-hand side of (2.37):

$$\left[\Box_1 + \left(\frac{mc}{\hbar}\right)^2\right] G_{12} \longrightarrow \left[\frac{-p \cdot p + (mc)^2}{\hbar^2} - \frac{i}{\hbar} p \cdot \partial_x + \frac{1}{4}\partial_x \cdot \partial_x\right] W(x, p) \,. \tag{2.43}$$

For the Minkowski product of four-vectors, we use the simplified notation $p^2 \equiv p \cdot p$ from now on. Consider now the physical meaning of the various terms occurring in this composite operator, acting on the Wigner transform of the Green function. The first term of order $1/\hbar^2$ is just the deviation of the 4-momentum argument in the Wigner function from its mass-shell value:

$$\hat{D} \equiv -p^2 + (mc)^2 \,. \tag{2.44}$$

The mixed term of order $1/\hbar$ is a transport operator:

$$\hat{T} \equiv p \cdot \partial_x = \frac{E}{c^2}\left(\frac{\partial}{\partial t} + \mathbf{v} \cdot \nabla\right) \,. \tag{2.45}$$

Here the velocity is classical, but relativistic, $\mathbf{v} = c^2\mathbf{p}/E$. In a comoving frame, the transport operator $\hat{T}$ is the time derivative times the rest mass: it is the classical parallel transporter along the worldline of a point particle. Reconstructing the quantum mechanical momentum operators defined by $\hat{P}_j = -i\hbar\partial_j$ for $j = 1, 2$, we can write the total operator in the suggestive form

$$\left[\Box_1 + \left(\frac{mc}{\hbar}\right)^2\right]G_{12} \longrightarrow \frac{1}{\hbar^2}\left[-\left(\frac{\hat{P}_1 + \hat{P}_2}{2} - p\right)^2 + (mc)^2\right]W\left(\frac{x_1 + x_2}{2}, p\right).$$

$$(2.46)$$

In the absence of interaction (zero self-energy), this expression and its equivalent form (2.43) must vanish. This condition we may call the 'quantum on-shell' requirement. It is obvious that it can deviate from the classical on-shell criterion $p^2 - (mc)^2 = 0$.

We note at this point that, for a spacetime translation invariant Green function, $G(x_1, x_2) = G^{\text{transl. inv}}(x_1 - x_2)$, its Wigner transform is independent of the common spacetime point $x = (x_1 + x_2)/2$, i.e., $W(x, p) = W^{\text{transl. inv}}(p)$. Fortunately, for such cases, the two on-shell criteria coincide:

$$\left[\Box_1 + \left(\frac{mc}{\hbar}\right)^2\right]G_{12}^{\text{transl. inv}} \longrightarrow \frac{1}{\hbar^2}\left[-p^2 + (mc)^2\right]W^{\text{transl. inv}}(p).\qquad(2.47)$$

The presence of interaction, causing a nontrivial self-energy contribution, changes this picture. In order to calculate the Wigner transform of the right-hand side of (2.37), besides the trivial contribution of the Dirac delta whose Wigner transform is the function identically equal to 1, we need to obtain the Wigner transform of the convolution term. This procedure is straightforward but a little tedious.

For an expression of the form $H(x_1, x_2) = \int dz \, F(x_1, z)G(z, x_2)$, the Wigner transform becomes

$$U(x, p) = \int d\xi \, e^{\frac{i}{\hbar}p\xi} \int dz F(x + \xi/2, z)G(z, x - \xi/2).\qquad(2.48)$$

The standard trick is to use a double representation of the integration variable

$$z = x + \varepsilon = (x - \xi/2) + (\varepsilon + \xi/2) = (x + \xi/2) + (\varepsilon - \xi/2).$$

Then the shift in the argument from the desired value is represented by a Taylor expansion, i.e., by an exponential of the 'shift times derivative' operator. In order to separate actions on the one and on the other factor under the integral, the idea is to change the arguments to primed ones and at the end of the derivation to restore $x' = x$. Following these steps, we arrive at

$$U(x, p) = \int d\varepsilon \int d\xi \, e^{\frac{i}{\hbar}p\xi} \left[e^{(\varepsilon + \xi/2)\partial_2} F(x_1, x_2)\right]\left[e^{(\varepsilon - \xi/2)\partial_1'} G(x_1', x_2')\right].\qquad(2.49)$$

We now replace the functions F and G by the Weyl transforms (inverses) of the corresponding Wigner functions:

$$F(x_1, x_2) = \int \frac{d^4 q}{(2\pi\hbar)^4} \, e^{-\frac{i}{\hbar}q(x_1 - x_2)} \, f(x, q) \,,$$

$$G(x_1', x_2') = \int \frac{d^4 r}{(2\pi\hbar)^4} \, e^{-\frac{i}{\hbar}r(x_1' - x_2')} \, g(x', r) \,, \qquad (2.50)$$

with $x_1 - x_2 = x_1' - x_2' = \xi$. Finally, it is useful to change the integration over the momentum variables q and r into one over their sum and difference variables, $q = \ell + m/2$ and $r = \ell - m/2$. Collecting together all the shift operators that express the phases in the exponentials, we obtain

$$U(x, p) = \int \hat{S}_m \hat{S}_\ell \hat{S}_x \hat{S}_{x'} f(x, p) g(x', p') \,. \qquad (2.51)$$

Here the integration is over ε, ξ, m, and ℓ with the proper phase space factors gathered together as a power of $2\pi\hbar$. The shift operators are as follows:

$$\hat{S}_m = e^{m\left(\frac{i}{\hbar}\varepsilon + \frac{1}{2}\partial_p - \frac{1}{2}\partial_{p'}\right)} \,, \qquad \hat{S}_\ell = e^{\ell\left(-\frac{i}{\hbar}\xi + \partial_p + \partial_{p'}\right)} \,,$$

$$\hat{S}_x = e^{\frac{1}{2}(\varepsilon + \xi/2)\partial_x} \,, \qquad \hat{S}_{x'} = e^{\frac{1}{2}(\varepsilon - \xi/2)\partial_{x'}} \,. \qquad (2.52)$$

The integration over ℓ and m ensures the actions

$$\varepsilon \to \frac{\hbar}{2i} \left(\partial_{p'} - \partial_p\right) \,, \qquad \text{and} \qquad \xi \to \frac{\hbar}{i} \left(\partial_{p'} + \partial_p\right) \,. \qquad (2.53)$$

The remaining integrations over ε and ξ leave us with the following total phase in the resulting shift operator:

$$\Delta = \frac{\hbar}{4i} \left(\partial_{p'} - \partial_p\right) \left(\partial_x + \partial_{x'}\right) + \frac{\hbar}{4i} \left(\partial_{p'} + \partial_p\right) \left(\partial_x - \partial_{x'}\right) = \frac{\hbar}{2i} \left[\partial_{p'}\partial_x - \partial_p\partial_{x'}\right] \,. \qquad (2.54)$$

All actions at the end have to be evaluated at $x' = x$ and $p' = p$. We conclude that

$$U(x, p) = \left. e^\Delta f(x, p) g(x', p') \right|_{x=x', p=p'} \,, \qquad (2.55)$$

and write the evolution equation in terms of Wigner transforms in the following shorthand notation:

$$\left[\frac{-p^2 + (mc)^2}{\hbar^2} - \frac{i}{\hbar} p \cdot \partial_x + \frac{1}{4}\partial_x^2\right] W = 1 + e^\Delta \, \Sigma W \approx 1 + \Sigma W + \frac{\hbar}{2i}\{\Sigma, W\} + \cdots \,. \qquad (2.56)$$

Here the last term in the approximation is the classical Poisson bracket.

It is noteworthy that the ansatz satisfying (2.56) to leading order in the $\hbar$ expansion, viz.,

$$W_{\text{LO}} = \frac{1}{\dfrac{p^2 - (mc)^2}{\hbar^2} + \Sigma} \, , \tag{2.57}$$

also satisfies the next to leading order operation

$$\frac{\hbar}{2i} \{\Sigma, W_{\text{LO}}\} = -\frac{i}{\hbar} p \cdot \partial_x W_{\text{LO}} \, . \tag{2.58}$$

Since the self-energy term Σ in this Wigner transform representation is in general complex, so is W_{LO}. The squared absolute value of the denominator in (2.57) is a resonance-like expression, having the resonance at a shifted on-shell mass value. The real part of the self-energy causes a shift in the rest mass due to interaction, while its imaginary part acts to produce the decay width of the quantum resonance.

2.3 Spectral Density and Its Statistical Parameters

Any Green function, or the corresponding Wigner function, evolves in time (and space) according to the interactions which 'dress' it. For thermal states, or more generally for equilibrium states, an additional symmetry applies: such states are time reversal invariant. Therefore it makes the physical background more transparent if Green and Wigner functions are viewed as a product of their equilibrium counterparts with a non-equilibrium factor developing in time. In the Fourier transformed picture, the time-independent equilibrium factor describes an energy-dependent 'occupation probability' for a given energy state. At finite temperature, with given density matrix in the traditional form $\hat{\rho} = \exp\left(-\hat{H}/T\right)/Z$, such thermal equilibrium factors become the Bose or Fermi distributions for free fields. As the interaction generates a nontrivial self-energy, it drives the Wigner function form of the propagators away from the free form. This can be described by assigning another factor, the spectral function, sensitive to off-mass-shell-ness due to quantum effects.

The spectral function $S_{AB}(\omega)$ is defined as the Fourier spectrum of a commutator correlation between two operators at different times, the commutator being taken in a given ensemble of quantum states described by the density operator $\hat{\rho}$:

$$S_{AB}(\omega) \equiv \int dt \, e^{-i\omega t} \, \text{Tr}\left(\hat{\rho}[\hat{A}(t), \hat{B}(0)]\right) \, . \tag{2.59}$$

This definition tacitly assumes a stationary t-independent density operator and a time-shift invariance property (conservation of total energy). The spectral function for Hermitian operators satisfies the relations

$$S_{AB}^*(\omega) = S_{B^\dagger A^\dagger}(\omega) \, , \qquad S_{AB}(-\omega) = -S_{BA}(\omega) \, ,$$

when the operators exhibit time-shift invariance. Analogously, one may consider the Wigner transform of the commutator as a generalized spectral function

$$S_{AB}(x, p) \equiv \int dq \; e^{\frac{i}{\hbar}pq} \langle \psi | \, [\hat{A}(x - q/2), \, \hat{B}(x + q/2)] \, | \psi \rangle \; . \qquad (2.60)$$

Whenever $\hat{A} = \hat{B}^\dagger$, the initial and final states defining the density matrix are changed by the same operator; but the respective shifts in spacetime are still opposite. It turns out that $S_{AB}(x, -p) = -S_{BA}(x, p)$ and $S_{AB}^*(x, p) = S_{B^\dagger A^\dagger}(x, p)$ for the Wigner commutator defined as above. On the other hand the Keldysh function is defined by the symmetric combination

$$i K_{AB}(x, p) \equiv \frac{1}{2} \int dq \; e^{\frac{i}{\hbar}pq} \langle \psi | \, \{\hat{A}(x - q/2), \, \hat{B}(x + q/2)\} \, | \psi \rangle \; , \qquad (2.61)$$

where $\{A, B\} = AB + BA$ denotes the anti-commutator. Its properties

$$i K_{AB}(x, -p) = i K_{BA}(x, p) \; , \qquad i K_{AB}^*(x, p) = -i K_{B^\dagger A^\dagger}(x, p) \; ,$$

ensure that, for the particular choice $B = A^\dagger$, both the Wigner commutator S and the Keldysh function iK are real functions of the phase space variables (x, p).

For a particular choice of operators, $B = A^\dagger = A = \Phi$, we obtain

$$S_{\Phi\Phi}(x, -p) = -S_{\Phi\Phi}(x, p) \; , \qquad S_{\Phi\Phi}^*(x, p) = S_{\Phi\Phi}(x, p) \; ,$$

i.e., the Wigner transformed spectral function is real-valued and an odd function of the energy and momentum. In analogy to the canonical equilibrium case, treated later, one can define an effective number of quanta, filling a phase space cell, by relating the Keldysh function and the spectral function in their Wigner transformed versions:

$$i K_{AB}(x, p) = \left[n_{AB}(x, p) + \frac{1}{2} \right] S_{AB}(x, p) \; . \qquad (2.62)$$

Noting that $(i K_{\Phi\Phi})^*(x, p) = i K_{\Phi\Phi}(x, p)$ and $S_{\Phi\Phi}(x, p)$ are real, one concludes that $n_{\Phi\Phi}^*(x, p) = n_{\Phi\Phi}(x, p)$ is also real. Furthermore, applying (2.62) for $A = B = \Phi = \Phi^\dagger$ (self-correlation of a Hermitian operator), one concludes that

$$n_{\Phi\Phi}(x, -p) + \frac{1}{2} = -\left[n_{\Phi\Phi}(x, p) + \frac{1}{2} \right] \; . \qquad (2.63)$$

This leads immediately to

$$\bar{n}_{\Phi\Phi}(x, p) \equiv -n_{\Phi\Phi}(x, -p) = n_{\Phi\Phi}(x, p) + 1 \; , \qquad (2.64)$$

the generalization of the bosonic Kubo–Martin–Schwinger (KMS) relation, confirming the Bose enhancement factor and the interpretation of antiparticles as negative energy particles 'propagating backwards in spacetime'.

For time-independent density operators, such as a thermal state defined by $\hat{\rho} = e^{-\beta \hat{H}}/Z$, only the time difference plays a role in the spectral function. When the time evolution of the test operators $\hat{A}$ and $\hat{B}$ is represented in the Heisenberg picture, their evolution driven by the same Hamiltonian as the one producing the thermal state, an important property of the spectral function can be derived. This Kubo–Martin–Schwinger (KMS) relation uses a cyclic permutation of expressions under the trace:

$$
\begin{aligned}
iG_{AB}(\omega) &= \int dt\, e^{-i\omega t}\, \mathrm{Tr}\left[e^{-\beta \hat{H}}\, e^{it\hat{H}/\hbar}\, \hat{A}(0)\, e^{-it\hat{H}/\hbar}\, \hat{B}(0) \right] \\
&= \int dt\, e^{-i\omega(t+i\hbar\beta)}\, e^{-\beta\hbar\omega}\left[e^{-\beta\hat{H}}\, e^{-i(t+i\hbar\beta)\hat{H}/\hbar}\, \hat{B}(0)\, e^{-i(t+i\hbar\beta)\hat{H}/\hbar}\, \hat{A}(0) \right] \\
&= e^{-\beta\hbar\omega}\, iG_{BA}(\omega)\,.
\end{aligned} \tag{2.65}
$$

From the commutator definition (2.59), we have $S_{AB} = iG_{AB} - iG_{BA}$. In particular applying the above general result to creation and annihilation operators $\hat{A} = a^{\dagger}$ and $\hat{B} = a$, respectively, one obtains the following relation for the number of quanta of energy $\hbar\omega$ in a plane wave:

$$
\langle \psi |\, a^{\dagger} a\, | \psi \rangle = e^{-\beta\hbar\omega}\, \langle \psi |\, a a^{\dagger}\, | \psi \rangle\,, \tag{2.66}
$$

which leads to

$$
n(\hbar\omega) S(\omega) = e^{-\beta\hbar\omega}\, [1 + n(\hbar\omega)]\, S(\omega)\,. \tag{2.67}
$$

The resolution of this relation for the thermal equilibrium distribution of quanta is the Bose–Einstein distribution at temperature $k_{\mathrm{B}} T = 1/\beta$:

$$
n(\hbar\omega) = \frac{1}{e^{\hbar\omega/k_{\mathrm{B}}T} - 1}\,. \tag{2.68}
$$

Let us now investigate the above expectation values and spectral functions for a general quantum state. It is interesting here to consider expectation values of differently ordered products of operators. The time ordering, denoted by $\mathscr{T}$, means an arrangement of operators according to their time arguments, e.g.,

$$
\mathscr{T} A(t) B(0) = \Theta(t) A(t) B(0) + \Theta(-t) B(0) A(t)\,. \tag{2.69}
$$

From now on we omit the hat symbol for operators, assuming that it will be clear from the context. Here, $\Theta(t)$ is the Heaviside function with value $+1$ for positive arguments and 0 for negative arguments. Its Fourier transform is

$$
\Theta(\omega) = \frac{i}{\omega + i\varepsilon}\,, \tag{2.70}
$$

where ε is a small positive number arbitrarily close to zero. With these preparations, we can define the following different-time field operator correlations, also called propagators:

$$
\begin{aligned}
iG_{AB}^{12}(t) &= \langle \psi | \, B(0)A(t) \, | \psi \rangle \, , \\
iG_{AB}^{21}(t) &= \langle \psi | \, A(t)B(0) \, | \psi \rangle \, , \\
iG_{AB}^{11}(t) &= \langle \psi | \, \mathcal{T} A(t)B(0) \, | \psi \rangle = \Theta(t)iG_{AB}^{21}(t) + \Theta(-t)iG_{AB}^{12}(t) \, , \\
iG_{AB}^{22}(t) &= \langle \psi | \, \mathcal{T}^* A(t)B(0) \, | \psi \rangle = \Theta(t)iG_{AB}^{12}(t) + \Theta(-t)iG_{AB}^{21}(t) \, . \quad (2.71)
\end{aligned}
$$

The expectation values in the generic quantum state $|\psi\rangle$ can be calculated with the help of the density matrix $\rho = |\psi\rangle\langle\psi|$.

These definitions are not all independent of each other, since we have the identity $\Theta(t) + \Theta(-t) = 1$. This implies

$$
G_{AB}^{11} + G_{AB}^{22} = G_{AB}^{12} + G_{AB}^{21} \, , \tag{2.72}
$$

which suggests that one linear combination of the four propagators defined above will vanish identically. Indeed, this is the retarded and advanced basis, formally defined by operators on two parallel paths in the complex ω plane (see the Keldysh formalism) [12]. Correlators between retarded and advanced operators are then collected in the formulas below:

$$
\begin{aligned}
iG_{AB}^{ra}(t) &= iG_{AB}^{11}(t) - iG_{AB}^{12}(t) = \Theta(t) \left(iG_{AB}^{21} - iG_{AB}^{12} \right) \, , \\
iG_{AB}^{ar}(t) &= iG_{AB}^{11}(t) - iG_{AB}^{21}(t) = -\Theta(-t) \left(iG_{AB}^{21} - iG_{AB}^{12} \right) \, , \\
iG_{AB}^{rr}(t) &= \frac{1}{2} \left[iG_{AB}^{21}(t) + iG_{AB}^{12}(t) \right] \, , \\
iG_{AB}^{aa}(t) &= 0 \, . \tag{2.73}
\end{aligned}
$$

Since $iG_{AB}^{ra}(t) = \Theta(t)S_{AB}(t)$ according to the above, its Fourier spectrum can be expressed in terms of the spectral function $S_{AB}(\omega)$ as follows:

$$
G_{AB}^{ra}(\omega) = \int \frac{d\omega'}{2\pi} \frac{S_{AB}(\omega')}{\omega - \omega' + i\varepsilon} \, , \tag{2.74}
$$

known as the Kramers–Kronig relation. Inverting this, the spectral function is just the discontinuity in the 'forward time' propagator in the complex frequency plane:

$$
S_{AB}(\omega) = \lim_{\varepsilon \to 0^+} \left[iG_{AB}^{ra}(\omega + i\varepsilon) - iG_{AB}^{ra}(\omega - i\varepsilon) \right] \, , \tag{2.75}
$$

which can easily be generalized to a wave four-vector $k = (\omega, \mathbf{k})$.

Finally, let us investigate some particular spectral functions more closely. Spectral functions belonging to a pair of operators, A and $A^\dagger$, in a canonical thermal state are real and include transition probabilities from one energy eigenstate to another. We consider

$$S_{AA^\dagger}(x) = \text{Tr}\left(e^{-\beta H}\left[A(x), A^\dagger(0)\right]\right) , \tag{2.76}$$

and rewrite the trace in terms of four-momentum eigenstates:

$$S_{AA^\dagger}(x) = \sum_P \langle P|\, e^{-\beta H}\left[A(x), A^\dagger(0)\right]|P\rangle . \tag{2.77}$$

Inserting now a complete system of states $|P'\rangle$ and taking into account the fact that four-momentum eigenstates are also eigenstates of the Hamiltonian, we obtain

$$S_{AA^\dagger}(x) = \sum_{PP'}\left(e^{-\beta E} - e^{-\beta E'}\right)\langle P|\,A(x)\,|P'\rangle\langle P'|\,A^\dagger(0)\,|P\rangle . \tag{2.78}$$

Finally, we use the fact that $A(x) = e^{iP\cdot x}\,A(0)\,e^{-iP\cdot x}$, and therefore

$$\langle P|\,A(x)\,|P'\rangle = e^{i(P-P')\cdot x}\,\langle P|\,A(0)\,|P'\rangle . \tag{2.79}$$

In this way, the transition matrix element between P and P' occurs with the initial operator $A(0)$ and its adjoint in a positive-definite product:

$$\langle P|\,A(0)\,|P'\rangle\langle P'|\,A^\dagger(0)\,|P\rangle = \left|\langle P|\,A(0)\,|P'\rangle\right|^2 , \tag{2.80}$$

and the above spectral function becomes

$$S_{AA^\dagger}(x) = \sum_{PP'}\left(e^{-\beta E} - e^{-\beta E'}\right)e^{i(P-P')\cdot x}\left|\langle P|\,A(0)\,|P'\rangle\right|^2 . \tag{2.81}$$

Finally, this result can be Fourier transformed and one arrives at

$$S_{AA^\dagger}(k) = \sum_{PP'}\left(e^{-\beta E} - e^{-\beta E'}\right)(2\pi)^4\delta^{(4)}(k + P - P')\left|\langle P|\,A(0)\,|P'\rangle\right|^2 . \tag{2.82}$$

This is nonzero only for operators $A(0)$ which have a non-vanishing matrix element between states with different energies.

References

1. E. Wigner, On the quantum correction for thermodynamic equilibrium. Phys. Rev. **40**, 749 (1932)
2. M. Hillery, R.F. O'Connell, M.O. Scully, E.P. Wigner, Distribution functions in physics: fundamentals. Phys. Rep. **106**, 121 (1984)
3. W.B. Case, Wigner functions and Weyl transforms for pedestrians. Am. J. Phys. **76**, 937 (2008)
4. W.P. Schleich, *Quantum Optics in Phase Space* (Wiley-VCH, Weinheim, 2001)

5. S. Varro, P. Adam, T.S. Biro, G.G. Barnafoldi, P. Levai, Wigner 111. Colorful and deep; scientific symposium, in *EPJ Web of Conferences*, vol. 78 (2014), pp. 00001–08002
6. M.A. Manko, V.I. Manko, Probability description and entropy of classical and quantum systems. Found. Phys. **41**, 330 (2011)
7. G. Baym, L.P. Kadanoff, Conservation laws and correlation functions. Phys. Rev. **124**, 287 (1961)
8. L.P. Kadanoff, G. Baym, *Quantum Statistical Mechanics* (Addison-Wesley, Boston, 1989)
9. K. Balzer, M. Bonitz, *Nonequilibrium Green's Functions Approach to Inhomogeneous Systems*. Springer Lecture Notes in Physics, vol. 867 (Springer, Berlin, 2013)
10. A.A. Vlasov, On vibrational properties of electron gas. J. Exp. Theor. Phys. **8**, 291 (1938)
11. A.A. Vlasov, *Many Particle Theory and Its Applications to Plasma* (Gordon and Breach, New York, 1961)
12. L.V. Keldysh, Zh. Eksp. Teor. Fiz. **47**, 1515 (1964) (in Russian); Sov. Phys. JETP **20**, 1018 (1965)

Chapter 3
Keldysh (Two-Time) Formalism

In previous chapters we dealt with the variance of the energy, its lower bound being established by quantum mechanical effects. We have shown that in a relativistic view this phenomenon converts into an off-mass-shell effect, and leads to a generalization of the probabilistic view of classical statistical physics to the use of a density operator and its spectral transform, the Wigner function. Yet, the Wigner function is just a particular view of the quantum field theoretical propagator.

On the other hand, any field theoretical treatment of the propagators unavoidably contains a part, called the spectral function, which reflects the quantum nature of entangled positive and negative frequency waves. In this chapter we deepen this discussion by re-formulating such statements in the framework of the Keldysh formalism, a mathematically tricky approach using two time directions in order to give a better description of causality (time order) on both the classical and the quantum levels. After putting forward arguments in favor of the Boltzmann–Gibbs density operator in a fixed temperature equilibrium, we go on to consider various path integral prescriptions and their relations to each other, and eventually give an overview of the concepts of perturbative expansion and renormalization procedures along the lines of constant physics.

3.1 Why the Boltzmann Distribution?

The problem of defining temperature naturally arises in quantum systems, too. Statistical properties can be included in quantum systems through the concept of the density matrix, which allows the description of a system even if there is information loss. The density matrix $\hat{\rho}$ delivers the expectation value of an operator $\hat{A}$ as

$$\langle \hat{A} \rangle_{\hat{\rho}} = \operatorname{Tr} \hat{\rho} \hat{A} . \tag{3.1}$$

© The Author(s), under exclusive licence to Springer Nature Switzerland AG 2019

T. S. Biró and A. Jakovác, *Emergence of Temperature in Examples and Related Nuisances in Field Theory*, SpringerBriefs in Physics,
https://doi.org/10.1007/978-3-030-11689-7_3

To ensure $\langle 1 \rangle = 1$, the density matrix has to have unit trace. Only Hermitian operators have real expectation values, so $\hat{\rho}$ must be Hermitian, too.

The general state of a quantum system is characterized by an arbitrary density matrix that satisfies the above conditions. If the system is time translation invariant, i.e., in a state where the operators $\hat{A}$ that are not explicitly time-dependent, viz., $\partial_t \hat{A} = 0$, have time-independent expectation values, then for all such operators,

$$0 = \frac{d}{dt} \langle \hat{A} \rangle_{\hat{\rho}} = \text{Tr}\, \hat{\rho} \frac{d}{dt} \hat{A} = \text{Tr}\, \hat{\rho} i[\hat{H}, \hat{A}] = -i\, \text{Tr}[\hat{H}, \hat{\rho}]\hat{A}\,, \qquad (3.2)$$

whence $[\hat{H}, \hat{\rho}] = 0$. This means that $\rho = \sum_E \rho(E) |E\rangle \langle E|$, where $\hat{H} |E\rangle = E |E\rangle$ are the energy eigenvectors. For other conserved quantities, $|E, N, \ldots\rangle$ should be the common eigenvectors. In this formula, $\rho(E)$ is not entirely fixed. The question is whether we can even further restrict the form of a stationary – later we also use the term equilibrium – density matrix.

Boltzmann observed that in most cases $\rho(E) \propto e^{-\beta E}$, where $\beta = 1/k_B T$ is the inverse temperature of the environment included in the density matrix. It was Gibbs, who first provided an argument in favor of this Boltzmann form. He based his argument on the additivity of the entropies and energies of independent subsystems, which leads to the extensivity of the energy and entropy in large systems. There also exist physical situations where it is useful to consider non-extensive concepts of thermodynamics containing non-additive entropy formulas [1, 2].

Following Gibbs' argument we divide the system into a small and a large subsystem. Then the total energy and entropy are written as

$$E_{\text{tot}} = E + E_L\,, \qquad S_{\text{tot}} = S + S_L\,, \qquad (3.3)$$

where the subscript L singles out the 'large' subsystem. Assuming independence, we have two separate relations, $S(E)$ and $S_L(E_L)$. The total entropy becomes

$$S_{\text{tot}} = S(E) + S_L(E_{\text{tot}} - E) \approx S(E) + S_L(E_{\text{tot}}) - \frac{\partial S_L}{\partial E} E + \cdots\,, \qquad (3.4)$$

whenever $E \ll E_{\text{tot}}$ holds. This means that the probability of observing the subsystem in a state with energy E is given by

$$\rho(E) = e^{S_L(E_{\text{tot}} - E) - S_{\text{tot}}(E_{\text{tot}})} \sim e^{-\beta E}\,, \qquad \beta = \frac{\partial S_L}{\partial E}\,. \qquad (3.5)$$

To generalize this line of thought to quantum systems, we should generalize the classical notion of entropy. The classical entropy is a characterization of the number of states satisfying certain conditions. To generalize it we need a *measure* in the Hilbert space. If we discretize the system in such a way that $\mathscr{H} \sim \mathbf{R}^N$, then it can be the usual volume measure in $\mathbf{R}^N$.

The energy functional in the Hilbert space is $\mathscr{E}(\psi) = \langle\,\psi|\hat{H}|\psi\,\rangle \in \mathbf{R}$, where $|\psi\rangle \in \mathscr{H}$. This is a continuous functional of the states, so its inverse $\mathscr{E}^{-1}$ applied to the open set $(]E, E + dE[)$ gives an open subset in $\mathscr{H}$. The energy-dependent entropy is defined via the area measure $\Omega(E)dE = |\mathscr{E}^{-1}(]E, E + dE[)|$ on the energy hypersurface, and $S(E) = \ln \Omega(E)$.

Once we have a definition for the entropy, we can proceed in the same way as in the classical case. Consider a quantum system in a closed volume V_{tot}. Let us single out a subsystem $V \subset V_{\text{tot}}$, and denote its complement by $\bar{V} = V_{\text{tot}} - V$. We also define a continuous function that restricts a global state to the subsystem $r_V : \psi \mapsto \psi_V = \Theta(x \in V)\psi(x)$, where $\Theta(x \in V) = 1$ if $x \in V$ and 0 otherwise. We define the subsystem energy by $\mathscr{E}_V(\psi) = \langle\,\psi_V|\hat{H}|\psi_V\,\rangle$ and the subsystem entropy by

$$S_V(E) = \ln\left(\frac{1}{dE}\,|\mathscr{E}_V^{-1}(]E, E + dE[)|\right). \tag{3.6}$$

The most important question here is this: if the state of the complete system evolves in time as $\psi(t)$, then what is the probability $P(\chi)$ that we find the subsystem in a given state $\psi_V(t) = \chi$ at a random time. Since the time evolution keeps the energy functional $\mathscr{E}(\psi(t)) = E_{\text{tot}}$ constant, then, if the time evolution is ergodic, $P(\chi)$ is proportional to the measure of the intersection of the subsets of the constraints: $P(\chi)dE \sim |\mathscr{E}^{-1}(]E_{\text{tot}}, E_{\text{tot}} + dE[) \cap r_V^{-1}(\chi)|$.

$P(\chi)$ can be assessed in a large enough subsystem. We then write the state as a sum $\psi = \psi_V + \psi_{\bar{V}}$. If the subsystem is large,

$$\langle\,\psi|\hat{H}|\psi\,\rangle \approx \langle\,\psi_V|\hat{H}|\psi_V\,\rangle + \langle\,\psi_{\bar{V}}|\hat{H}|\psi_{\bar{V}}\,\rangle. \tag{3.7}$$

The difference from the exact expression is $2\text{Re}\langle\,\psi_{\bar{V}}|\hat{H}|\psi_V\,\rangle$, which we refer to as the interface energy. The above condition therefore requires the interface energy to be negligible compared with the 'bulk energy'. This can usually be achieved by sending $|V|, |\bar{V}| \to \infty$ while $|V|/|\bar{V}| \to 0$.

If the above approximation is valid, then the condition $\psi_V = \chi$ implies that the first term on the right-hand side of (3.7) is fixed. We denote it by $E = \langle\,\psi_V|\hat{H}|\psi_V\,\rangle$. Therefore the intersection of constraints simply means that the second term should have an energy $E_{\text{tot}} - E$, whence $P(\chi)dE \sim |\mathscr{E}_{\bar{V}}^{-1}(]E_{\text{tot}} - E, E_{\text{tot}} - E + dE[)|$. This leads to

$$P(\chi) \sim e^{S_{\bar{V}}(E_{\text{tot}} - E)} \sim e^{-\beta E}, \qquad \beta = \frac{\partial S_{\bar{V}}(E)}{\partial E}. \tag{3.8}$$

We may draw some lessons from the quantum analysis. First of all the E in the exponent is the expectation value of the Hamiltonian on a restricted state confined to a subvolume, and not the subsystem energy level. In fact the subsystem is an open system whose Hamiltonian is not Hermitian, so its energy eigenvalues are no longer real.

A second remark is that both the classical and the quantum version of the derivation of the Boltzmann factor strongly depend on assumptions about the interface energy. One can imagine a system where the interface energy always remains an important factor. Then the Boltzmann–Gibbs assumption is no longer valid [1].

3.2   Two-Time Path Integration

If we start from a pure state $|\psi\rangle$, then the expectation value of a time-local measurement represented by a Hermitian operator $\hat{A}$ reads $\langle\psi|\hat{A}|\psi\rangle$. However, if the measurement lasts for a longer time, we inevitably measure also a time average

$$\langle\,\hat{A}\,\rangle = \frac{1}{\delta t} \int\limits_{t-\delta t/2}^{t+\delta t/2} dt'\, w(t')\langle\psi,t'|\hat{A}|\psi,t'\rangle\,, \tag{3.9}$$

where $w(t')$ is a possible weight for the measurement. This average can be written

$$\langle\,\hat{A}\,\rangle = \mathrm{Tr}\,\hat{\rho}\hat{A}\,, \qquad \hat{\rho} = \frac{1}{\delta t} \int\limits_{t-\delta t/2}^{t+\delta t/2} dt'\, w(t')\,\big|\psi,t'\big\rangle\big\langle\psi,t'\big|\,. \tag{3.10}$$

Further averaging does not spoil the above way of taking the expectation value, thanks to the linearity in $\hat{\rho}$. The weights are usually chosen to ensure $\mathrm{Tr}\,\hat{\rho} = 1$.

The time evolution of the density matrix in units such that $\hbar = c = 1$ is given by $\hat{\rho}(t) = e^{-i\hat{H}t}\hat{\rho}e^{i\hat{H}t}$, where t stands for $t - t_0$ and t_0 is the initial time. The time evolution can also be shifted to the measurement operator, since $\mathrm{Tr}\,e^{-i\hat{H}t}\hat{\rho}e^{i\hat{H}t}\hat{A} = \mathrm{Tr}\,\hat{\rho}\hat{A}(t)$, where $\hat{A}(t) = e^{i\hat{H}t}\hat{A}e^{-i\hat{H}t}$. In this (Heisenberg) picture it is easy to write down the correlation between measurements performed at different times $t_1 \ldots t_n$:

$$\langle\,\hat{A}_1(t_1)\ldots\hat{A}_\ell(t_\ell)\,\rangle = \mathrm{Tr}\,\rho\hat{A}_1(t_1)\ldots\hat{A}_\ell(t_\ell)\,. \tag{3.11}$$

Writing the explicit time evolution into this formula, we obtain for the right-hand side

$$\mathrm{Tr}\,\rho e^{i\hat{H}t_1}\,\hat{A}_1 e^{-i\hat{H}(t_1-t_2)}\,\hat{A}_2\ldots e^{-i\hat{H}(t_{n-1}-t_n)}\,\hat{A}_n e^{-i\hat{H}t_n}\,. \tag{3.12}$$

To have a practical formula for evaluating this trace, we have to represent the time evolution in some way. The *path integral representation* uses a complete basis $|n\rangle$ $(n \in N)$, i.e.,

$$\mathbf{1} = \sum_n |n\rangle\,\langle n|\,, \tag{3.13}$$

and inserts unit operators to give

$$e^{-i\hat{H}t} = \prod_{a=1}^{N} e^{-i\hat{H}dt} = \sum_{n_0\ldots n_N} |n_0\rangle \left[\prod_{a=0}^{N-1} \langle n_a|e^{-i\hat{H}dt}|n_{a+1}\rangle \right] \langle n_N| . \tag{3.14}$$

Then we introduce the notation $H(n_a, n_{a+1})$ by requiring

$$\frac{\langle n_a|e^{-i\hat{H}dt}|n_{a+1}\rangle}{\langle n_a|n_{a+1}\rangle} = e^{-iH(n_a,n_{a+1})dt} , \tag{3.15}$$

to leading order in a dt expansion. In general, the 'classical' Hamiltonian H introduced in this way depends on the chosen interval dt, but in the limit $dt \to 0$ and assuming that the time evolution is differentiable, we expect this dependence to disappear. In this limit,

$$H(n_a, n_{a+1}) = \frac{\langle n_a|\hat{H}|n_{a+1}\rangle}{\langle n_a|n_{a+1}\rangle} . \tag{3.16}$$

We can also introduce a formal Lagrangian as

$$L(n_a, n_{a+1}) = i\frac{\ln\langle n_a|n_{a+1}\rangle}{dt} - H(n_a, n_{a+1}) . \tag{3.17}$$

With this notation we can represent the time evolution operator as

$$e^{-i\hat{H}(t-t_0)} = \sum_{n_0\ldots n_N} |n_0\rangle\, e^{i\int_{t_0}^{t} L(n_{t'},n_{t'+dt'})dt'} \langle n_N| , \tag{3.18}$$

where $t' = t_0 + n_a dt$.

At this point, the following remarks are in order:

- $T_{n_a,n_{a+1}} = e^{iL(n_a,n_{a+1})dt}$ corresponds to the transfer matrix of statistical physics.
- In some basis $\langle n_a| n_{a+1}\rangle$ may vanish, giving a zero in the denominator. However, this never happens in a coherent state basis, where there is always a nonzero overlap between two coherent states, or if the x and p bases are used alternatingly along the path. In a strict sense, the limit $dt \to 0$ is not generally sensible, since the logarithmic term involving $|n_a\rangle$ and $|n_{a+1}\rangle$, two mutually independent states in a quantum path, may be large even for a short time lapse.

Now we can write down the expectation value of a generic operator [see (3.11)]. First we rewrite it to show the explicit time dependence:

$$\langle\, \hat{A}_1(t_1)\ldots\hat{A}_\ell(t_\ell)\,\rangle = \text{Tr}\,\rho e^{i\hat{H}(t_1-t_0)}\hat{A}_1 e^{-i\hat{H}(t_1-t_2)}\ldots\hat{A}_\ell(t_\ell)e^{-i\hat{H}(t_\ell-t_0)}. \tag{3.19}$$

We can then insert the representation (3.18) into the above form, but we should take care about the indexing of the basis states: in the representation of $e^{i\hat{H}(t_i - t_{i+1})}$ for different indices i, we refer to different states, so we cannot simply index them by the time $t_0 + n_a dt$ alone. For better book-keeping, we introduce a *contour time* τ that grows monotonically in the subsequent representations. In real time, $\tau \mapsto t$ goes as $t_0 \to t_\ell \to t_{\ell-1} \to \ldots t_2 \to t_1 \to t_0$. This is thus a closed time path, hence the name closed time path formalism (CTP). The path integral naturally provides an expression that is contour-time-ordered, i.e., the operators appear in order of their increasing contour time from right to left. If we assume that the representing states $|n\rangle$ are eigenvectors of $\hat{A}_i$, and denote $\langle n_a | \hat{A}_i | n_a \rangle = A_i(\tau_i)$, then we obtain

$$\langle T_\tau \hat{A}_1(\tau_1) \ldots \hat{A}_\ell(\tau_\ell) \rangle = \sum_{n_i \ldots n_f} \langle n_i | \hat{\rho} | n_f \rangle e^{i \int_{\tau_i}^{\tau_f} L(n_\tau, n_{\tau + d\tau}) d\tau} A_1(\tau_1) \ldots A_\ell(\tau_\ell) .$$

$$(3.20)$$

This is the most general path integral representation.

To simplify the situation, we may wish to compute expectation values of operators where $t_0 < t_\ell < t_{\ell-1} < \cdots < t_x > t_{x-1} > t_{x-2} > \cdots > t_1 > t_0$, so τ grows until t_x, then decreases (evidently, for $\ell = 2$, this is always the case). In this case, the closed path can be divided into two segments, one strictly increasing, the other monotonically decreasing in time. Note that we can insert $1 = e^{-i\hat{H}(t_x - t_y)} e^{-i\hat{H}(t_y - t_x)}$ in the evolution, so the path can always be extended from $t_x \to t_{x-1}$ to $t_x \to t_y \to t_x \to t_{x-1}$, and the contours of the two segments may run as $C_1 : t_0 \to \infty$ and $C_2 : \infty \to t_0$. Due to the monotonicity, we can parameterize these contours with physical time. In this representation, all time-dependent operators have an extra index referring to their respective contours: $A^{(1)}(t)$ and $A^{(2)}(t)$. The complete Lagrangian can also be written as a sum of $L^{(1)}$ and $L^{(2)}$, but in the second case the change in the integration variable results in an extra minus sign ($\int_{C_2} d\tau = -\int_{t_0}^\infty dt$). Thus we have

$$L = L^{(1)} - L^{(2)} .$$

$$(3.21)$$

With this we arrive at the result

$$\langle \hat{A}_1(t_1) \ldots \hat{A}_\ell(t_\ell) \rangle = \sum_{n_i \ldots n_f} \langle n_i | \hat{\rho} | n_f \rangle e^{i \int_{t_0}^\infty L(n_t, n_{t+dt}) dt} A_1^{(a_1)}(t_1) \ldots A_\ell^{(a_\ell)}(t_\ell) , \quad (3.22)$$

where $a_i = 1$ or 2.

The density matrix for the Boltzmann distribution looks like a time translation operator with $t = -i\beta$. Therefore the path integral representation is the same as (3.18):

$$\hat{\rho}_{\text{local}} = \frac{1}{Z} \sum_{n_0 \ldots n_N} |n_0\rangle e^{-\int_0^\beta L_E(n_\tau, n_{\tau + d\tau}) d\tau} \langle n_N| ,$$

$$(3.23)$$

where

$$L_E(n_a, n_{a+1}) = -\frac{\ln\langle n_a | n_{a+1}\rangle}{dt} + H_{local}(n_a, n_{a+1}) - \mu_i N_{local,i}(n_a, n_{a+1}) \quad (3.24)$$

is a 'Euclidean' Lagrangian. Formally,

$$L_E(\tau) = -L_{local}(t_0 - i\tau) - \mu_i N_{local,i}(n_a, n_{a+1}) .$$

This representation can be added as a new contour segment to the previous case. Thus we obtain

$$\frac{1}{Z} \operatorname{Tr} e^{-\beta(\hat{H} - \mu \hat{N})} \hat{A}_1(t_1) \dots \hat{A}_\ell(t_\ell) = \frac{1}{Z} \sum_{n_i \dots n_f = n_i} e^{iS} A_1^{(a_1)}(t_1) \dots A_\ell^{(a_\ell)}(t_\ell) , \quad (3.25)$$

where now $a_i = 1, 2,$ or 3, and

$$iS = i \int_{t_0}^{\infty} \left[L^{(1)} - L^{(2)} \right] - \int_0^{\beta} L_E . \quad (3.26)$$

As a consequence of the trace, $|n_f\rangle = |n_i\rangle$.

In the quantum mechanical case, the representation of unity is most easily taken as

$$\mathbf{1} = \int dp \, dq \, |q\rangle \langle q | p \rangle \langle p| . \quad (3.27)$$

Then the formal Lagrangian (3.17) reads

$$L_E(q_a, p_{a+1}) = -\frac{\ln\langle q_a | p_{a+1}\rangle \langle p_{a+1} | q_{a+1}\rangle}{dt} + H_{local}(q_a, p_{a+1}) - \mu_i N_{local}(q_a, p_{a+1}) . \quad (3.28)$$

Taking into account the fact that $\langle q | p \rangle = e^{ipq}$, we have in the continuum limit

$$L_E(q, p) = -ip\dot{q} + H_{local}(q, p) - \mu_i N_{local}(q, p) , \qquad \dot{q} = \frac{q_{a+1} - q_a}{dt} . \quad (3.29)$$

If $H_{local}(q, p) = p^2/2 + V(q)$, while N_{local} does not depend on p, then

$$\int dp \, e^{ip\dot{q} - \frac{1}{2}p^2} = \int dp \, e^{-\frac{1}{2}(p - i\dot{q})^2} e^{-\frac{1}{2}\dot{q}^2} \sim e^{-\frac{1}{2}\dot{q}^2} . \quad (3.30)$$

In this case, $L_E(q)$ will really be the Euclidean Lagrangian.

For bosonic field theories, we may pursue the same method. There, the fundamental field operator is a function of time and space $\hat{\Phi}(t, x)$. From the canonical commutation relations, we know that $[\hat{\Phi}(t, x), \hat{\Phi}(t, y)] = 0$, indicating that there is a common eigensystem of operators $\hat{\Phi}(t, x)$ for all x :

$$\exists |\varphi\rangle \text{ such that } \forall x , \quad \hat{\Phi}(t, x) |\varphi\rangle = \varphi(t, x) |\varphi\rangle , \quad (3.31)$$

and with a similar definition we may find $|\pi\rangle$ as a common eigenvector of $\hat{\Pi}(t, \boldsymbol{x})$, the canonically conjugate fields.

We can now discretize the space, i.e., we consider operators at the positions $\boldsymbol{x}_n = \sum_{i=1}^{d} n_i a_i \boldsymbol{e}_i$, where d is the dimension of the space, $n \in \boldsymbol{Z}^d$, a_i are quantities with a certain length scale (the lattice spacing), and $\boldsymbol{e}_a$ is the unit vector pointing in the direction a. We also use the notation $\varphi_n = \varphi(\boldsymbol{x}_n)$ and $\pi_n = \pi(\boldsymbol{x}_n)$. We represent unity by

$$\mathbf{1} = \int \prod_n \mathrm{d}\varphi_n \mathrm{d}\pi_n \; |\varphi\rangle \, \langle \varphi | \pi \rangle \, \langle \pi | \; . \tag{3.32}$$

This formula is fully equivalent to the multidimensional quantum mechanical case, so all previous formulas can be applied without change. In particular, the correlation functions can be calculated as

$$\frac{1}{Z} \operatorname{Tr} e^{-\beta(\hat{H} - \mu \hat{N})} \hat{\Phi}_1(x_1) \ldots \hat{\Phi}_\ell(x_\ell) = \frac{1}{Z} \int \mathscr{D}\varphi e^{\mathrm{i}S[\varphi]} \varphi_1^{(a_1)}(x_1) \ldots \varphi_\ell^{(a_\ell)}(x_\ell) \;, \tag{3.33}$$

where $a_i = 1, 2$, or 3,

$$\mathrm{i}S = \mathrm{i} \int \mathrm{d}^4 x \left[\mathscr{L}^{(1)} - \mathscr{L}^{(2)} \right] - \int \mathrm{d}^3 x \int_0^\beta \mathscr{L}_E \;, \tag{3.34}$$

and the integration measure reads

$$\int \mathscr{D}\varphi = \prod_{i,n} \mathrm{d}\varphi_n^{(1)}(t_i) \mathrm{d}\varphi_n^{(2)}(t_i) \prod_{i,n} \mathrm{d}\varphi_n^{(3)}(\tau_i) \;. \tag{3.35}$$

For fermionic systems the situation is somewhat more complicated. The interested reader should consult [3]. In this case the field variables do not commute, but anti-commute, so there is no common eigensystem for fields at different spatial positions. The solution to this problem is to introduce a Hilbert space over the Grassmann algebra, which heuristically can be thought as made up of anticommuting real variables. With this trick, the path integral can be formulated. Fortunately, fermions always appear in bilinear forms in renormalizable theories, and then their path integral can be performed exactly. As a result we have a pure bosonic action with a somewhat complicated Lagrangian containing nonlocal terms.

In equilibrium, a simplification can be made. Since the density matrix is time translation invariant, we can push the initial time to $-\infty$. In this case the 3rd contour and the $t = 0$ moment are infinitely far from each other, and no influence can propagate between the two. This leads to the *factorization* property, which means that we can deal either with the 3rd contour (in the imaginary time or Matsubara formalism) or with the contour 1–2 (in the real time or Keldysh formalism).

This means that among all the two-point functions it is sufficient to use only the 11, 12, 21, 22, and 33 components discussed earlier. The 33 component is usually defined without the imaginary unit factor i:

$$G_{33}(\tau) = \langle T_\tau \Phi(\tau)\Phi(0) \rangle . \tag{3.36}$$

Since the imaginary time evolution goes with the same analytic formula as the real time evolution, we conclude that

$$G_{33}(\tau) = \Theta(\tau)G_{21}(-i\tau) + \Theta(-\tau)G_{12}(-i\tau) . \tag{3.37}$$

In particular, in the scalar field theory case and for $\tau > 0$, this leads to

$$G_{33}(\tau) = \int \frac{d\omega}{2\pi} e^{-\omega\tau} G_{21}(\omega) = \int \frac{d\omega}{2\pi} \frac{e^{\omega(\beta-\tau)}}{e^{\beta\omega}-1} S(\omega) = \int_0^\infty \frac{d\omega}{2\pi} \frac{\cosh\omega\left(\dfrac{\beta}{2}-\tau\right)}{\sinh\dfrac{\beta}{2}\omega} S(\omega) .$$
$$\tag{3.38}$$

In a similar way, we find $G_{33}(-\tau) = G_{33}(\tau)$. This result also shows that $G_{33}(\tau) = G_{33}(\beta - \tau)$. From these two we find

$$G_{33}(\tau + \beta) = G_{33}(\tau) . \tag{3.39}$$

Although the imaginary time argument of the propagator is in the range $[-\beta, \beta]$ (since that comes from the difference between two imaginary time arguments less than β), we can extend the G_{33} propagator to all τ values, using the same formula, as a periodic function of τ with period β.

Thus the Fourier transform of the imaginary time propagator is defined as

$$G_{33}(\omega = 2\pi nT) = \int_0^\beta d\tau\, G_{33}(\tau) e^{-i\tau\omega} . \tag{3.40}$$

From the above relations it follows that

$$G_{33}(\omega) = \int \frac{d\omega'}{2\pi} G_{21}(\omega') \int_0^\beta d\omega\, e^{-(\omega'+i\omega)\tau} = \int \frac{d\omega'}{2\pi} \frac{S(\omega')}{\omega' + i\omega} = -G^{ra}(-i\omega) . \tag{3.41}$$

3.3 Perturbative Expansion

Now we have an exact tool to calculate any expectation value with an arbitrary density matrix, and for any time dependence. But this is only a theoretical possibility, since the resulting integrals are much too complex to be calculated. From a practical point of view, therefore, we have to make compromises.

One possibility is that we assume that the action can be split into two parts, viz.,

$$S[\varphi] = S_0[\varphi] + S_{\text{int}}[\varphi] , \tag{3.42}$$

where keeping only S_0 provides a solvable system (within which all expectation values can be calculated). Then we may write

$$\langle \hat{A} \rangle = \frac{1}{Z} \int \mathscr{D}\varphi \, e^{iS_0[\varphi]+iS_{\text{int}}[\varphi]} A[\varphi] = \frac{Z_0}{Z} \langle e^{iS_{\text{int}}} \hat{A} \rangle_0 , \tag{3.43}$$

where the zero subscript means taking the expectation value in the S_0 system:

$$\langle \hat{B} \rangle_0 = \frac{1}{Z_0} \int \mathscr{D}\varphi \, e^{iS_0[\varphi]} B[\varphi] , \qquad Z_0 = \int \mathscr{D}\varphi \, e^{iS_0[\varphi]} . \tag{3.44}$$

Thus we have to compute an expectation value in the solvable system. In practice, one expands the exponent in a power series

$$\langle \hat{A} \rangle = \frac{Z_0}{Z} \sum_{n=0}^{\infty} \frac{1}{n!} \langle (iS_{\text{int}}[\varphi])^n A[\varphi] \rangle_0 = \frac{1}{Z} \sum_{n=0}^{\infty} \frac{1}{n!} \int \mathscr{D}\varphi \, e^{iS_0[\varphi]} (iS_{\text{int}}[\varphi])^n A[\varphi] . \tag{3.45}$$

In this way, whenever this series converges, we can only calculate expectation values to arbitrary precision by computing expectation values in the original, solvable system.

The basis of this perturbative approach is the knowledge of expectation values in some particular system. This is usually chosen to be an interaction-free system, where the action is at most quadratic in the fields. For real-valued bosonic fields, this means considering an action

$$S[\varphi] = \frac{1}{2} \sum_{a,b} \varphi_a K_{ab} \varphi_b , \tag{3.46}$$

where a and b run over all spacetime points in a given discretization, and K_{ab} is the kernel of the quadratic form. For example, we may have

$$S = \frac{1}{2} \int d^4x \, \varphi(x)(-\partial^2 - m^2)\varphi(x) , \tag{3.47}$$

where the Fourier-transformed kernel is $K(p) = p^2 - m^2$, if we calculate in the continuum limit of vanishing lattice spacing $a \to 0$.

In this Gaussian case we can compute all expectation values explicitly. The operator equation of motion is

$$(\partial^2 + m^2)\hat{\Phi} = 0 , \tag{3.48}$$

so we know the time evolution of the Fourier modes:

$$\hat{\Phi}(t, k) = \hat{\Phi}(k) \cos \omega_k t + \hat{\Pi}(k) \frac{\sin \omega_k t}{\omega_k} , \tag{3.49}$$

with $\omega_k^2 = k^2 + m^2$. The spectral function reads

$$S(t, k) = \langle \, [\hat{\Phi}(t, k), \hat{\Phi}(0, x = 0)] \, \rangle = -i \frac{\sin \omega_k t}{\omega_k} , \tag{3.50}$$

using the canonical commutation relation in the form $[\hat{\Phi}(0, x = 0), \hat{\Pi}(0, k)] = i$. This leads to the on-mass-shell spectral function

$$S(k) = 2\pi \, \text{sgn}(k_0) \delta(k^2 - m^2) . \tag{3.51}$$

If we know the spectral function, we know all the free propagators, too. Using the Kramers–Kronig relation, we find the retarded propagator

$$G^{ra}(k) = \frac{1}{k^2 - m^2} \bigg|_{k_0 \to k_0 + i\varepsilon} . \tag{3.52}$$

The 12 and 21 propagators come from the KMS relation

$$G_{12}(k) = n(k_0) S(k) , \qquad G_{21}(k) = \big[1 + n(k_0)\big] S(k) . \tag{3.53}$$

In particular,

$$G_{12}(x = 0) = \int \frac{d^3 k}{(2\pi)^3} \left[\frac{1}{2} + n(\omega_k) \right] . \tag{3.54}$$

From (3.41), the 33 propagator reads

$$G_{33}(k) = \frac{1}{k^2 + m^2} . \tag{3.55}$$

We still have to compute the 'zero point function', i.e.,

$$Z = \text{Tr} \, e^{-\beta H} = \int \prod_a d\varphi_a e^{-\frac{1}{2}\varphi_a K_{ab}\varphi_b} = (\det K)^{-1/2} . \tag{3.56}$$

Form this, $\ln Z = -\frac{1}{2} \text{Tr} \ln K$. Using the thermodynamic relation $\ln Z = -\beta F$, where F is the free energy, we obtain

$$F = \frac{T}{2} \ln \det K = \frac{T}{2} \text{Tr} \ln K . \tag{3.57}$$

Here we view the matrix K in the 3-momentum eigenstate representation:

$$K(p) = \sum_{\mathbf{p},\ell} \left[p_0 - \omega(\mathbf{p})\right] |\mathbf{p}, \ell\rangle \langle\mathbf{p}, \ell| \; . \tag{3.58}$$

In Fourier space, we use the discrete momenta $k_n = 2\pi n/L$ in each direction, with the distance $\delta k = 2\pi/L$. In the continuous phase space limit, the trace is equivalent to an integral

$$\mathrm{Tr}(\ldots) = \beta V \int \frac{\mathrm{d}^4 p}{(2\pi)^4} \; \ldots \; . \tag{3.59}$$

For the free energy density $f = F/V$, we thus obtain

$$f = \frac{1}{2} \int \frac{\mathrm{d}^4 p}{(2\pi)^4} \; \ln K(p) \; . \tag{3.60}$$

For this calculation, only the third contour contributes. We also apply a trick, and compute f for $K(p) + \lambda\mathbf{1}$ instead of $K(p)$ alone. We then take the limit $\lambda \to 0$ at the very end of the calculation. This addition replaces $p_0 \to p_0 - \lambda$ in (3.58). We consider the derivative with respect to λ and then use the fact that

$$\frac{\mathrm{d}}{\mathrm{d}\lambda} \ln K(p + \lambda) = K^{-1}(p + \lambda) = G(p + \lambda)$$

to deduce that

$$\frac{\partial f}{\partial \lambda} = \frac{T}{2} \int \frac{\mathrm{d}^4 p}{(2\pi)^4} \, G_{33}(p, \lambda) = \mathrm{i}G_{12}(\tau = 0, \lambda) = \int \frac{\mathrm{d}^3 k}{(2\pi)^3} \left[n(\omega_k + \lambda) + \frac{1}{2}\right] . \tag{3.61}$$

Here we observed that, at the origin of the complex time plane, where the real time and imaginary time contours meet, $G_{33}(\tau = 0)$ is the same as $\mathrm{i}G_{12}(\tau = 0)$ [see (3.37)]. For the last step, we have used (3.54). The integrand can be written in the equivalent form

$$n(\omega + \lambda) + \frac{1}{2} = \frac{1}{2} \coth \frac{\beta(\omega + \lambda)}{2} \; . \tag{3.62}$$

Its primitive function with respect to λ is

$$\int \mathrm{d}\lambda \, \frac{1}{2} \coth \left[\frac{\beta(\omega + \lambda)}{2}\right] = \frac{1}{\beta} \ln \sinh \left[\frac{\beta(\omega + \lambda)}{2}\right] . \tag{3.63}$$

The definite integral from $\lambda = 0$ to $\lambda = \infty$ cancels the ω-dependent zero point motion factor and, collecting all infinite contributions in $f(\infty)$, we obtain

$$f(\lambda = 0) = T \int \frac{\mathrm{d}^3 k}{(2\pi)^3} \; \ln(1 - \mathrm{e}^{-\beta\omega_k}) \; . \tag{3.64}$$

Higher point functions can be generated by the generator functional

$$Z[J] = \int \mathcal{D}\varphi \, e^{iS+J\varphi} , \qquad (3.65)$$

where J contains the 1, 2, and/or 3 contours. We have

$$iG_{i_1 \ldots i_n}(x_1, \ldots, x_n) \equiv \langle T\Phi_{i_1}(x_1) \ldots \Phi_{i_n}(x_n) \rangle = \frac{1}{Z} \frac{\partial^n Z[J]}{\partial J_{i_1}(x_1) \ldots \partial J_{i_n}(x_n)} \bigg|_{J=0} . \qquad (3.66)$$

In the Gaussian case, the generator functional can be evaluated, but care must be taken over the boundary conditions. The result is

$$Z[J] = Z e^{\frac{i}{2} JGJ} , \qquad (3.67)$$

where G denotes the propagators. This implies *Wick's theorem*

$$\langle T\Phi_{i_1}(x_1) \ldots \Phi_{i_n}(x_n) \rangle = \sum_{\mathcal{P} \text{ pairings}} \prod_{<ab> \in \mathcal{P}} iG_{i_a i_b}(x_a - x_b) . \qquad (3.68)$$

Since we can now calculate all correlation functions in the original, unperturbed theory, we have to apply (3.45) to do perturbation theory. This is the direct method, and all the properties of perturbation theory follow from this formula. At the same time, perturbation theory involves many subtleties that would themselves fill a book; here we just mention the results.

Equation (3.45) involves evaluation of free correlators which, by Wick's theorem, are products of propagators, so the perturbative series is a sum in which each term is a product of propagators. These terms can be represented graphically (*Feynman diagrams*) by the following rules for evaluating $iG_{i_1 \ldots i_n}(k_1, \ldots, k_n)$ in Fourier space:

- Each external field is represented by a point labeled with a Keldysh (contour) index i_a and momentum k_a, from which a single line can emerge.
- If the interaction part of the action contains a term like

$$S_{\text{int}} \sim \int d^d x \, \lambda \Phi_i^m = \int \frac{d^d p_1}{(2\pi)^d} \cdots \frac{d^d p_m}{(2\pi)^d} \lambda (2\pi)^d \delta(p_1 + \cdots + p_m) \Phi_i(p_1) \ldots \Phi_i(p_m) , \qquad (3.69)$$

where i is the Keldysh index, then it is represented by a point (*vertex*) labeled by i and p_a, from which m lines emerge. The contribution of the vertex is

$$\lambda \int \frac{d^d p_1}{(2\pi)^d} \cdots \frac{d^d p_m}{(2\pi)^d} (2\pi)^d \delta(p_1 + \cdots + p_m) .$$

- If we have drawn all the external points and vertices corresponding to the given order of perturbation theory [see (3.45)], then we have connected all the points with lines in all possible ways, forming graphs. If a point labeled by i and p is connected by a line to a point labeled by j and q, then its contribution is

$$iG_{ij}(p)(2\pi)^d\delta(p+q) \; .$$

- The value of the graph/diagram is the product of the contributions of the lines and the vertices. After taking into account the Dirac-delta constraints, we are left with some integrals over the propagators and a possible momentum dependence from the vertices. If there remain integrals over ℓ, we say that the diagram is an ℓ-loop diagram.

3.4   Is There a Spacetime Continuum?

Although it is formally equivalent to the operator formalism, the path integral representation is in fact more fundamental than the operator-level description of quantum theory. In quantum mechanics, in the operator formalism, it is easy to say that the operators act on the wave functions, and in extreme situations, that the wave function may be concentrated at a single point (point particle). But this is not fully correct, as can be demonstrated by the following paradox: take the position operator $\hat{q}$ and its normalized eigenvectors $|q\rangle$. Then we have $\hat{q}\,|q\rangle = q\,|q\rangle$ and $\langle q|q\rangle = 1$ (at finite volume). Now let us take the expectation value of the commutator $[\hat{p}, \hat{q}]$ in such a state. We can do the calculation in two ways. First, using the canonical commutation relations, second, expanding the commutator. We have

$$\langle q|[\hat{q}, \hat{p}]|q\rangle = i\hbar = \langle q|\hat{q}\hat{p} - \hat{p}\hat{q}|q\rangle = q(\langle q|\hat{p}|q\rangle - \langle q|\hat{p}|q\rangle) = 0 \; . \quad (3.70)$$

The lesson is that the assumption of normalized eigenvectors and canonical commutation relations actually contradict each other, even at finite volume.

A pragmatic approach to the resolution of this paradox defines our model on a discrete space and time. In this case the Hilbert space remains finite dimensional, and all operators will have discrete spectra. For the definition of the quantum world, we can then use the path integral, which inherently contains discreteness by construction. In this sense the path integral is a *natural way* to define a quantum system.

So let us work on a spacetime mesh with lattice spacing a, although in fact other discretization procedures lead to the same conclusion. The discreteness can be thought of as a regulator, whose only important property is its finiteness, but a can be arbitrarily small. When we want to have physical results, we take the limit $a \to 0^+$. We speak about the 'continuum theory', if this limit exists and is unique.

But, as it turns out by computing the perturbative expressions, or trying to evaluate the path integral explicitly, this is not true. Starting from any interacting theory other than a quadratic one, most of the correlation functions diverge when we take the limit $a \to 0^+$. For a long time, this was considered as a proof that the quantum field theory itself was ill-defined. The situation is in a certain sense simpler, and more exciting at the same time.

In order to explore the mathematics behind this, we consider the path integral as a mathematical operation P that assigns a real number to a generalized polynomial of the fields:

$$P \;:\; \varPhi(x_1)\varPhi(x_2)\ldots\varPhi(x_n) \longmapsto \langle\, T\hat{\varPhi}(x_1)\hat{\varPhi}(x_2)\ldots\hat{\varPhi}(x_n)\,\rangle \in \boldsymbol{R}\,. \qquad (3.71)$$

It is also linear, and thus P is an element of the dual of the generalized polynomials, which are the (generalized) distributions:

$$P \in \mathscr{G}P\,. \qquad (3.72)$$

The path integral P depends on some parameters: one is the regulator a, i.e., the lattice spacing, the other is the action itself. From a practical point of view, the action is characterized by the coupling constants of the terms appearing in it, using an adequate operator basis

$$S[\varPhi] = \sum_n \lambda_n O_n[\varPhi]\,. \qquad (3.73)$$

Thus P depends on a and $\{\lambda\}$ as parameters.

In this language the observation that in the continuum limit most expectation values are divergent can be reformulated so that at fixed $\{\lambda\}$ the limit $a \to 0^+$ leads to a divergent distribution. But this is not too surprising a result. Consider for example the distribution

$$p(\varepsilon, \eta; x) = \frac{\eta}{x^2 + \varepsilon^2}\,. \qquad (3.74)$$

This is a kernel function that can act as a distribution on any finite support C_∞ functions according to

$$p(f) = \int\limits_{-\infty}^{\infty} \mathrm{d}x\, f(x) p(\varepsilon, \eta; x)\,. \qquad (3.75)$$

It is not difficult to see that this distribution is the analogue of the path integral P with two parameters.

The 'continuum limit' of the path integrals, which comes from $a \to 0$ and $\{\lambda\}$ finite, corresponds to the $\varepsilon \to 0$ limit while η is finite. In this case, p as a function almost everywhere goes to

$$\lim_{\varepsilon \to 0} p(\varepsilon, \eta; x) = \frac{\eta}{x^2}\,, \qquad (3.76)$$

which is a divergent distribution: it assigns divergent results to non-divergent test functions. Its naive continuum limit is ill-defined.

But p is a two-parameter set, and if we set $\eta \to 0$, then its value becomes zero:

$$\lim_{\eta \to 0} p(\varepsilon, \eta; x) = 0\,. \qquad (3.77)$$

Therefore the double limit $\lim_{\varepsilon\to 0, \eta\to 0}$ does not exist. The result depends on the particular way the limits are taken. If, for example, we take the limit on the line $\eta = r\varepsilon$, we obtain an 'ordinary' distribution:

$$\lim_{\varepsilon\to 0,\, \eta=r\varepsilon} p(\varepsilon, \eta; x) = \lim_{\varepsilon\to 0} \frac{r\varepsilon}{x^2 + \varepsilon^2} = r\pi\delta(x) . \tag{3.78}$$

The lesson is that we obtain a sensible continuum limit along a line in the (ε, η) plane. We can say that the line is defined so that the value resulting from the action of the distribution is fixed at

$$p(f) = r\pi f(0) . \tag{3.79}$$

Thus the continuum limit is controlled by a 'physical' quantity, the value of the distribution on a fixed test function, e.g., $f(x)$ with $f(0) = 1$.

The knowledge gained by the study of this example can be taken over to the more complicated case of the path integral. We know that

$$\lim_{a\to 0} P(a, \{\lambda\}) \longrightarrow \text{divergent distribution} . \tag{3.80}$$

We also find that, if the coupling constants go to infinity, then the phase of the integral oscillates wildly, and

$$\lim_{\{\lambda\}\to\infty} P(a, \{\lambda\}) = 0 . \tag{3.81}$$

This situation very much resembles the example studied previously. This allows us to define a meaningful continuum limit along a path, $\lambda_i(\varepsilon)$, in the $(\varepsilon, \{\lambda\})$ space. This special path is determined so that the values of some physical quantities remain fixed. Hence the name 'line of constant physics' (LCP). There is no 'continuum limit' in general; each such LCP defines its own continuum.

This also means that the continuum limit is not simply $a \to 0$ with fixed $\{\lambda\}$, but is rather defined by the LCP:

$$\text{continuum limit}: \quad a \to 0 , \quad \{\lambda\} = \{\lambda(a)\} . \tag{3.82}$$

As a consequence, the continuum limit yields cutoff-dependent, or running couplings. This is also known as 'the' *renormalization* procedure.

References

1. C. Tsallis, *Introduction to Nonextensive Statistical Mechanics* (Springer, Berlin, 2009)
2. T.S. Biro, *Is There a Temperature?*. Fundamental Theories of Physics, vol. 1014 (Springer, Berlin, 2011)
3. A. Jakovac, A. Patkos, *Resummation and Renormalization in Effective Theories of Particle Physics*. Lecture Notes in Physics, vol. 912 (Springer, Berlin, 2016)

Chapter 4
Linear Response

In this chapter we continue our study of the quantum field theoretical description of the behavior of systems responding dynamically to external perturbations. The dynamical response in the linear approximation reflects some universal features which may help us to understand how a temperature emerges.

Although the formula (3.22) can be used to compute expectation values with an arbitrary density matrix, we have so far only considered equilibrium systems. Nonequilibrium systems are much harder to discuss, for several reasons.

Let us start with the physical difficulty. A general nonequilibrium state is complicated. In many cases it can only be described with infinitely many parameters. This also means that we do not know what the density matrix will be in a given experiment; it will be impossible to measure all the details. Conversely, we cannot prepare a given density matrix with all the necessary details. *This means that, even if we could calculate the path integral, it is useless in any experimental situation.* Nevertheless, after a certain time we expect the wild quantum oscillations to be damped, and we expect the system to approach a steady or equilibrium state. For most operators their expectation values are presumably very close to the equilibrium ones, their damping rate being large. Only a few operators are expected to survive the approach to equilibrium for a long time, and even for them an expansion around the equilibrium value should be a good approximation. This is the basic idea behind the *linear response*, the topic of the present section.

The second difficulty is of a technical nature. In naive perturbation theory starting from an arbitrary density matrix, *secular terms* appear that grow in time, rendering the perturbative series less and less convergent [1, 2]. The easiest way to demonstrate this is to consider the free propagator in time

$$G_{21}(\boldsymbol{k}, t) = \frac{e^{-i\omega_k t}}{2\omega_k} , \qquad (4.1)$$

where $\omega_k^2 = k^2 + m_0^2$. If for some reason the mass starts to deviate from the initial value, then the frequency changes to

© The Author(s), under exclusive licence to Springer Nature Switzerland AG 2019
T. S. Biró and A. Jakovác, *Emergence of Temperature in Examples and Related Nuisances in Field Theory*, SpringerBriefs in Physics,
https://doi.org/10.1007/978-3-030-11689-7_4

$$\omega_k' = \sqrt{\omega_k^2 + \delta m^2} \approx \omega_k + \frac{\delta m^2}{2\omega_k} + \mathcal{O}(\delta m^4) \,, \tag{4.2}$$

where δm^2 is proportional to λ, the coupling to the force that changes the mass. Rewriting it into the propagator, we find

$$e^{-i\omega_k' t} = e^{-i\omega_k t}\left[1 - i\frac{\delta m^2}{2\omega_k}t + \mathcal{O}(\delta m^4)\right] \,. \tag{4.3}$$

Since the first correction is $\mathcal{O}(\lambda)$, the perturbative correction grows linearly with time. This example shows that, even for the simplest cases, a naive perturbative approach cannot be applied. We have to do *resummations* before obtaining physically reasonable results.

4.1   Calculating the Linear Response

Let us prepare a non-equilibrium system in which the dynamic processes can be studied. We follow the strategy of a real experimental physicist, who goes to the equilibrium laboratory in the morning, applies some changes there, then does the measurements. This means that we start from an equilibrium system at $t = -\infty$, apply some external force on it, and study the change in the expectation value of some operators. For simplicity, the external force will be characterized by a single operator $\hat{F}(x)$, and its strength will be $\zeta(x)$, a spacetime dependent c-number function. The action of the system is therefore expressed as

$$S_{\text{tot}} = S - \zeta(x)F(x) \,, \tag{4.4}$$

where S denotes the action of the unperturbed system. The expectation value of an operator $\hat{O}(x)$ then reads

$$\langle \hat{O}(x) \rangle = \int \mathscr{D}\Phi_1 \mathscr{D}\Phi_2 \, e^{iS[\Phi_1] - iS[\Phi_2] - i\zeta F_1 + i\zeta F_2} O_1(x) \,, \tag{4.5}$$

where the 1 and 2 indices refer to the Keldysh modes.[1] We expand this expression in a power series with respect to ζ, obtaining to linear order

$$\langle \hat{O}(x) \rangle = \langle \hat{O}(x) \rangle_{\text{eq}} - i \int d^4x' \langle T_c O_1(x)\left[F_1(x') - F_2(x')\right] \rangle_{\text{eq}} \zeta(x') \,, \tag{4.6}$$

where the equilibrium expectation values are taken at $\zeta = 0$. According to the Keldysh rules, the contour-ordered expectation value is causally retarded, i.e.,

---

[1] The expression in the exponential is also called the influence functional.

$$\langle T_c \hat{O}_1(x)[F_1(x') - F_2(x')] \rangle = \Theta(t - t')\langle [\hat{O}(x), F(x')] \rangle . \qquad (4.7)$$

The right-hand side here describes the retarded propagator for the O and F operator pair, G^{ra}_{OF}. So we can write the change in the expectation value as

$$\langle \hat{O}(x) \rangle - \langle \hat{O}(x) \rangle_{eq} = -i \int d^4x' \Theta(t - t')\langle [\hat{O}(x), F(x')] \rangle_{eq} \zeta(x') + \mathcal{O}(\zeta^2) .$$
$$(4.8)$$

In Fourier space, using the spacetime translation invariance of the equilibrium, we obtain

$$\langle \hat{O}(p) \rangle - \langle \hat{O}(p) \rangle_{eq} = -iG^{ra}_{OF}(p)\zeta(p) + \mathcal{O}(\zeta^2) . \qquad (4.9)$$

If necessary, we can also go to second order in ζ. The formulas are somewhat more involved, but the logic of the calculation is the same as before. For the second order term, we obtain an expression containing a double commutator at equilibrium:

$$(-i)^2 \int d^4x' d^4x'' \Theta(t - t')\Theta(t' - t'')\langle [[\hat{O}(x), F(x')], F(x'')] \rangle_{eq} \zeta(x')\zeta(x'') .$$
$$(4.10)$$

4.2 Response to an External Field

An interesting application of the linear response theory is given when the external force stems from an external field strength, for example the electric field $\mathbf{E}$. The power, describing the dissipation of energy due to Ohmic resistance, is given by

$$P = \partial_t H = - \int d^3x \, \mathbf{E}(t, \mathbf{x}) \mathbf{j}(t, \mathbf{x}) . \qquad (4.11)$$

We integrate this expression with respect to time to obtain the change in the Hamiltonian

$$\delta H(t) = - \int_{-\infty}^{t} dt' \int d^3x \, \mathbf{E}(t', \mathbf{x}) \mathbf{j}(t', \mathbf{x}) . \qquad (4.12)$$

Consequently, the change in the action reads as

$$S_{tot} = S + \int_{-\infty}^{\infty} dt \int_{-\infty}^{t} dt' \int d^3x \, \mathbf{E}(t', \mathbf{x}) \mathbf{j}(t', \mathbf{x}) . \qquad (4.13)$$

We now use (4.8) and write to first order

$$\langle \hat{O}(x) \rangle - \langle \hat{O}(x) \rangle_{eq} = \int d^3x' \int_{-\infty}^{t} dt' \int_{-\infty}^{t'} dt'' \langle i[\hat{O}(x), j_i(x'')] \rangle_{eq} E_i(x'') . \quad (4.14)$$

It is worth expressing this formula in Fourier space. Using translation invariance in equilibrium, the commutator expectation value yields a spectral function:

$$\langle [\hat{O}(x), j_i(x'')] \rangle_{eq} = S_{Oj_i}(x - x'') , \quad (4.15)$$

and we obtain in Fourier space the change

$$\langle \hat{O}(p) \rangle - \langle \hat{O}(p) \rangle_{eq} = i \int_{-\infty}^{\infty} dt\, e^{ip_0 t} \int_{-\infty}^{t} dt' \int_{-\infty}^{t'} dt''\, S_{Oj_i}(t - t'', p) E_i(t', p) .$$

$$(4.16)$$

Substituting the time Fourier transform into this expression, the right-hand side becomes

$$i \int \frac{d\omega_1 d\omega_2}{(2\pi)^2} S_{Oj_i}(\omega_1, p) E_i(\omega_1, p) \int_{-\infty}^{\infty} dt \int_{-\infty}^{t} dt' \int_{-\infty}^{t'} dt''\, e^{ip_0 t - i\omega_1(t-t'') - i\omega_2 t''} . \quad (4.17)$$

The time integrals are straightforward to perform, and we arrive at

$$\langle \hat{O}(p) \rangle - \langle \hat{O}(p) \rangle_{eq} = -i \int \frac{d\omega}{2\pi} \frac{S_{Oj_i}(\omega, p)}{(p_0 - \omega + i\varepsilon)^2} E_i(p) . \quad (4.18)$$

Here we can proceed in two different ways. The first leads to the observation that

$$\int \frac{d\omega}{2\pi} \frac{S_{Oj_i}(\omega, p)}{(p_0 - \omega + i\varepsilon)^2} = -\partial_{p_0} \int \frac{d\omega}{2\pi} \frac{S_{Oj_i}(\omega, p)}{p_0 - \omega + i\varepsilon} = -\partial_{p_0} G^{ra}_{Oj_i}(p) , \quad (4.19)$$

using the Kramers–Kronig relation. This results in

$$\langle \hat{O}(p) \rangle - \langle \hat{O}(p) \rangle_{eq} = \left[i\partial_{p_0} G^{ra}_{Oj_i}(p) \right] E_i(p) . \quad (4.20)$$

To proceed the other way requires some assumptions. First, we assume that $O = j_i$, so we look for the response for the same operator as the one that has caused the disturbance. In this case, $S_{j_i j_i}(\omega)$ is an odd function in ω, and $\langle j_i \rangle_{qe} = 0$. Second, we assume that the applied electric field is constant; at least constant on the space and time scales of the microscopic processes. Then $E(p) = E\delta(p)$, and so

$$\langle j_i \rangle = -i \int \frac{d\omega}{2\pi} \frac{S_{j_i j_i}(\omega, 0)}{(\omega - i\varepsilon)^2} E_i . \quad (4.21)$$

Using the fact that the spectral function is odd in ω, we may write

$$
\int \frac{d\omega}{2\pi} \frac{S_{j_i j_i}(\omega, 0)}{(\omega - i\varepsilon)^2} = \frac{1}{2} \int \frac{d\omega}{2\pi} S_{j_i j_i}(\omega, 0) \left[\frac{1}{(\omega - i\varepsilon)^2} - \frac{1}{(\omega + i\varepsilon)^2} \right]
$$

$$
= \frac{1}{2} \int \frac{d\omega}{2\pi} S_{j_i j_i}(\omega, 0) \frac{4i\varepsilon\omega}{(\varepsilon^2 + \omega^2)^2}
$$

$$
\xrightarrow{\varepsilon \to 0} \lim_{\omega \to 0} \frac{i S_{j_i j_i}(\omega, 0)}{\omega} , \tag{4.22}
$$

since $2\varepsilon/(\omega^2 + \varepsilon^2) \to 2\pi \delta(\omega)$ and $\omega/(\varepsilon^2 + \omega^2) \to 1/\omega$. So finally we arrive at

$$
\langle j_i \rangle = \lim_{\omega \to 0} \frac{S_{j_i j_i}(\omega, 0)}{\omega} E_i . \tag{4.23}
$$

Thus the current is proportional to the external field strength and the proportionality constant is the conductivity, viz.,

$$
\sigma = \lim_{\omega \to 0} \frac{S_{j_i j_i}(\omega, 0)}{\omega} , \tag{4.24}
$$

which is the *Kubo formula* [3–5].

4.3 Hydrodynamics and Field Theory

The treatment of hydrodynamics in field theory is logically similar to the above discussion, but it needs special care in some other places. In this section we briefly review how the transport coefficients, generally used in hydrodynamics, can be calculated from operator expectation values.

Hydrodynamics deals with problems where the deviation from equilibrium appears due to application of a flow. Furthermore, the temperature follows a non-uniform pattern. How can we assign operators to model these effects? Our discussion here is based on [6, 7].

In the case of a flow we first realize that the Hamiltonian, which is the zeroth component of the energy–momentum four-vector, is not Lorentz invariant. We need to obtain its transformation properties by changing to a finite velocity observer, characterized by its four-velocity u_μ. We use the fact that $u_\mu P^\mu$ is Lorentz invariant. Introducing $\bar{u}_\mu = (1, 0, 0, 0)$, we write

$$
H_0 = \bar{u}_\mu P_0^\mu = u_\mu P^\mu = \bar{u}_\mu P^\mu + (u - \bar{u})_\mu P^\mu = H + (u - \bar{u})_\mu P^\mu . \tag{4.25}
$$

By generalizing this thought, we can write

$$H = H_0 - \int d^3x \big[u(x) - \bar{u}\big]_\mu T^{0\mu}(x) \,, \tag{4.26}$$

where $T^{\mu\nu}$ is the energy–momentum tensor, so that $T^{0\mu}$ is the energy four-current density.

To mimic temperature changes, we should modify the system energy in a time dependent way. Thus $H \to (1 - \delta c)H$ will be the new Hamiltonian. Up to first order, the energy change is $\delta E = -\delta c E$, thus

$$\delta c = -\frac{\delta E}{E} = -\frac{\delta T}{T} = \frac{\delta\beta}{\beta} \,, \tag{4.27}$$

where we have used $E \sim T$.[2]

These two effects together result in the following change in the Hamiltonian

$$\delta H = -\int d^3x \big[U(x) - \bar{u}\big]_\mu T^{0\mu}(x) \,, \tag{4.28}$$

where $U_\mu = (1 + \delta c)u_\mu$.

A constant four-velocity does not mean non-equilibrium, so the real measure of the deviation from equilibrium is the derivative of U. To express δH in terms of this quantity, we first take its time derivative:

$$\partial_0 \delta H = -\int d^3x \big[\partial_0 U_\mu T^{0\mu} + (U - \bar{u})_\mu \partial_0 T^{0\mu}\big] \,. \tag{4.29}$$

Since $T^{\mu\nu}$ is conserved, $\partial_\mu T^{\mu\nu} = 0$, and after partial integration, we have[3]

$$\partial_0 \delta H = -\int d^3x \big[\partial_0 U_\mu T^{0\mu} - (U - \bar{u})_\mu \partial_i T^{i\mu}\big] = -\int d^3x\, \partial_\nu U_\mu T^{\nu\mu} \,. \tag{4.30}$$

Since $T^{\mu\nu}$ is symmetric, we may write

$$\delta H = -\int_{-\infty}^{t} dt' \int d^3x\, [\partial U]_{\nu\mu} T^{\nu\mu} \,, \tag{4.31}$$

with the symmetrized expression $[\partial U]_{\nu\mu} = \frac{1}{2}(\partial_\nu U_\mu + \partial_\mu U_\nu)$.

To proceed, we have to use the transformation properties of the system to ensure that the transport is diagonal. Since a particular four-vector is singled out, namely the velocity of the flow, Lorentz invariance reduces to a simple rotation invariance. Thus we may define the following projectors $\Pi^{(a)}$:

---

[2]This assumption does not hold for a material medium containing massive particles.
[3]The $T^{i\mu}\partial_i \bar{u}_\mu$ term vanishes.

$$\Pi^{(1)}_{\mu\nu;\rho\sigma} = u_\mu u_\rho u_\nu u_\sigma \,,$$

$$\Pi^{(2)}_{\mu\nu;\rho\sigma} = \frac{1}{3}\Delta_{\mu\nu}\Delta_{\rho\sigma} \,,$$

$$\Pi^{(3)}_{\mu\nu;\rho\sigma} = \Delta_{\mu\rho}u_\nu u_\sigma \,,$$

$$\Pi^{(4)}_{\mu\nu;\rho\sigma} = \frac{1}{2}\left(\Delta_{\mu\rho}\Delta_{\nu\sigma} + \Delta_{\nu\rho}\Delta_{\mu\sigma} - \frac{2}{3}\Delta_{\mu\nu}\Delta_{\rho\sigma}\right) \,, \tag{4.32}$$

with the transverse projector

$$\Delta_{\mu\nu} = g_{\mu\nu} - u_\mu u_\nu \,, \quad \text{satisfying} \quad \Delta_\mu{}^\sigma \Delta_\sigma{}^\nu = \Delta_\mu{}^\nu \,, \quad \Delta_\mu{}^\nu u_\nu = 0 \,. \tag{4.33}$$

It is easy to check that they form a complete set of projectors:

$$\Pi^{(a)\alpha\beta}_{\mu\nu}\Pi^{(b)\rho\sigma}_{\alpha\beta} = \delta^{ab}\Pi^{(a)\rho\sigma}_{\mu\nu} \,. \tag{4.34}$$

Unity can be decomposed as

$$g_{\mu\rho}g_{\nu\sigma} = \sum_a \Pi^{(a)}_{\mu\nu;\rho\sigma} = \sum_a \Pi^{(a)}_{\mu\nu;\alpha\beta}\Pi^{(a)\alpha\beta}_{\rho\sigma} \,. \tag{4.35}$$

The traces of the projectors are

$$\Pi^{(a)\mu\nu}_{\mu\nu} = T_a \,, \quad T_1 = 1 \,, \quad T_2 = 1 \,, \quad T_3 = 3 \,, \quad T_4 = 5 \,, \tag{4.36}$$

whence $\sum_a T_a = 10$, the total number of independent components of a symmetric 4×4 matrix.

We decompose the energy–momentum tensor in terms of these projectors to obtain

$$T^{\mu\nu} = \varepsilon u^\mu u^\nu - p\Delta^{\mu\nu} + Q^\mu u^\nu + Q^\nu u^\mu + \pi^{\mu\nu} \,, \tag{4.37}$$

where

$$\varepsilon = u_\rho u_\sigma T^{\rho\sigma} \,, \quad p = -\frac{1}{3}\Delta_{\rho\sigma}T^{\rho\sigma} \,, \quad Q_\mu = \Delta_{\mu\rho}u_\sigma T^{\rho\sigma} \,, \quad \pi_{\mu\nu} = \Pi^{(4)}_{\mu\nu;\rho\sigma}T^{\rho\sigma} \,. \tag{4.38}$$

These terms have direct physical meaning: ε is the energy density, p is the pressure, Q_μ the heat current, and $\pi_{\mu\nu}$ the stress tensor. The expression for δH in (4.31) can therefore be written as

$$T^{\mu\nu}[\partial U]_{\mu\nu} = \varepsilon(u^\mu u^\nu \partial_\mu U_\nu) - p(\Delta^{\mu\nu}\partial_\mu U_\nu) + Q^\mu(u^\nu \partial_\nu U_\mu + u^\nu \partial_\mu U_\nu)$$

$$+ \frac{1}{2}\pi^{\mu\nu}\left[\Delta_{\mu\rho}\Delta_{\nu\sigma}(\partial^\rho U^\sigma + \partial^\sigma U^\rho) - \frac{2}{3}\Delta_{\mu\nu}\Delta_{\rho\sigma}\partial^\rho U^\sigma\right] \,. \tag{4.39}$$

As a next step we should ensure that all terms are first order corrections. We can use the fact that $u^2 = 1$ and $\Delta u = 0$ in the calculation. We also use $\delta c = \delta \beta / \beta$ and apply the zeroth order energy–momentum equation (Euler equation) to arrive at the following evolution equation for the expectation value of the energy–momentum tensor:

$$u_\mu \partial_\nu \langle T^{\mu\nu} \rangle = 0 = u_\mu \partial^\mu \langle \varepsilon \rangle + \langle \varepsilon + p \rangle \partial^\mu u_\mu . \tag{4.40}$$

We also specify our system at rest with $u_\mu = \delta_{0\mu}$ and $\Delta_{\mu\nu} \to -\delta_{ij}$. Then, after some algebra, we find

$$\delta H = -\int_{-\infty}^{t} dt' \int d^3 x \left[p' \partial_i u_i + T^{0i} (Du_i + \beta^{-1} \partial_i \beta) + \frac{1}{2} \pi^{ij} \left(\partial_i u_j + \partial_j u_i - \frac{2}{3} \delta_{ij} \partial_i u^i \right) \right] ,$$
$$\tag{4.41}$$

where $Du_i = u^j \partial_j u_i$ and

$$p' = p - \left. \frac{\partial p}{\partial \varepsilon} \right|_{eq} \varepsilon . \tag{4.42}$$

Now these terms have different transformation properties, so their cross terms in the expectation values are zero. Therefore we obtain the following changes to expectation values in the low-frequency limit expressed by the corresponding Kubo factors [see (4.24)] based on spectral functions:

$$\delta \langle p' \rangle = \lim_{\omega \to 0} \frac{S_{p',p'}(\omega)}{\omega} \partial^i u_i ,$$

$$\delta \langle T^{0i} \rangle = \lim_{\omega \to 0} \frac{S_{T^{0i}, T_{0i}}(\omega)}{\omega} (Du^i + \beta^{-1} \partial^i \beta) ,$$

$$\delta \langle \pi^{ij} \rangle = \lim_{\omega \to 0} \frac{S_{\pi^{ij}, \pi_{ij}}(\omega)}{\omega} \frac{1}{2} \left(\partial^i u^j + \partial^j u^i - \frac{2}{3} \delta^{ij} \partial_m u^m \right) . \tag{4.43}$$

The coefficients correspond to the following terms in phenomenological hydrodynamics:

$$\delta \langle p' \rangle = \zeta \partial_i u_i ,$$
$$\delta \langle T_{0i} \rangle = -\kappa (Du_i + \beta^{-1} \partial_i \beta) ,$$
$$\delta \langle \pi_{ij} \rangle = \eta \left(\partial_i u_j + \partial_j u_i - \frac{2}{3} \delta_{ij} \partial_m u_m \right) , \tag{4.44}$$

these being the bulk viscosity ζ, the thermal conductivity κ, and the shear viscosity η. Finally, we arrive at the following Kubo formulas:

$$\zeta = \lim_{\omega \to 0} \frac{\langle [p', p'] \rangle(\omega)}{\omega} \, ,$$

$$\kappa = \lim_{\omega \to 0} \frac{\langle [T_{01}, T_{01}] \rangle(\omega)}{\omega} = \frac{1}{3} \lim_{\omega \to 0} \frac{\langle [T_{0i}, T_{0i}] \rangle(\omega)}{\omega} \, ,$$

$$\eta = \frac{1}{2} \lim_{\omega \to 0} \frac{\langle [\pi_{12}, \pi_{12}] \rangle(\omega)}{\omega} = \frac{1}{10} \lim_{\omega \to 0} \frac{\langle [\pi_{ij}, \pi_{ij}] \rangle(\omega)}{\omega} \, , \qquad (4.45)$$

where the signs in the last equation are valid due to the fact that the different components have the same expectation value in the same multiplet. Finally, we recall that

$$\langle [A, A] \rangle(\omega) = S_{AA}(\omega) = \int_{-\infty}^{+\infty} \langle [A(t), A(0)] \rangle \, \mathrm{e}^{\mathrm{i}\omega t} \, \mathrm{d}t \qquad (4.46)$$

are the corresponding spectral functions.

References

1. D. Boyanovski, H.J. de Vega, Dynamical renormalization group approach to relaxation in quantum field theory. Ann. Phys. **307**, 335 (2003)
2. S. Iso, H. Ohta, T. Suyama, Secular terms in Dyson series to all orders of perturbation. Prog. Theor. Exp. Phys. **2018/8**, 083A01 (2018)
3. M.S. Green, Markoff random processes and the statistical mechanics of time-dependent phenomena, II. Irreversible processes in fluids. J. Chem. Phys. **22**, 398 (1954)
4. R. Kubo, Statistical-mechanical theory of irreversible processes, I. General theory and single applications to magnetic and conduction problems. J. Phys. Soc. Jpn. **12**, 570 (1957)
5. R. Zwanzig, Time-correlation functions and transport coefficients in statistical mechanics. Ann. Rev. Phys. Chem. **16**, 67 (1965)
6. R. Horsley, W. Schonmaker, Quantum field theories out of thermal equilibrium, 1. general considerations. Nucl. Phys. B **280**, 716 (1987)
7. A. Hosoya, M. Shakagami, M. Takao, Nonequilibrium thermodynamics in field theory: transport coefficients. Ann. Phys. **154**, 229 (1984)

Chapter 5
Fluctuation, Dissipation, and Non-Boltzmann Energy Distributions

After reviewing the field theoretical description of the conventional thermal equilibrium and studying the linear relaxation to this equilibrium, it is interesting to consider problems arising in small systems not satisfying the pre-requisites for the Gibbs–Boltzmann treatment. In this chapter, we discuss general questions of thermal ensembles and the stochastic dynamics picture of a practically unknown environment, and at the end we sketch briefly the way the Keldysh formalism incorporates physical noise into its descriptive framework.

5.1 Emergence of Thermal Ensembles

There is a basic difference between the descriptions of deterministic and stochastic behaviors of dynamical systems and the resulting observables in classical and quantum physics. It lies in the introduction of the random aspect, in the form of variables and effects whose actual values cannot be sharply calculated or even defined, and for which only their mediated effect on averages can be derived. This differs from and comes in addition to the quantum 'uncertainty' we have already touched upon in previous chapters. A thermal state in quantum physics—and hence also in quantum field theory—is characterized by a density operator which cannot be reconstructed as a pure state density operator, i.e., $\hat{\rho} \neq |\Psi\rangle \langle \Psi|$. Logically, this could not happen if we had full quantum information on a closed system, but it frequently happens when considering the restriction of such systems to some smaller part. Formally, in thermal field theory, this problem is 'resolved' by doubling the Hilbert space in terms of Fock states [1–3], and deriving a Bogoliubov transformed Fock vacuum as the 'thermal vacuum'. However, this formal treatment gives us no clue how Nature masters this trick.

How do canonical thermal distributions with probability $e^{-\beta E}/Z$ emerge for a state with energy E? Since this is the basis for constructing the density operator in the form $\hat{\rho} = e^{-\beta \hat{H}}/Z$ for quantum systems, it is fundamentally important to answer this question in terms of general thermodynamics before dealing with applications. The key is to measure a subsystem of a whole, large, energetically closed system, as

© The Author(s), under exclusive licence to Springer Nature Switzerland AG 2019
T. S. Biró and A. Jakovác, *Emergence of Temperature in Examples and Related Nuisances in Field Theory*, SpringerBriefs in Physics,
https://doi.org/10.1007/978-3-030-11689-7_5

already touched upon in Sect. 3.1. The hand-waving mention of the 'thermodynam-ical limit' usually reminds us that the systems are large, and the Boltzmann–Gibbs exponential formula is then the main result. But is it possible to quantify the devi-ations caused by the finiteness of the total, or by the too small size of the observed parts? Indeed it is, in some simple cases.

When considering a subsystem of a bigger total system, the wave function is restricted to a smaller finite range. Boundary conditions are fixed, and there is com-binatorically a huge number of ways to continue a description of a given subsystem state into the whole. This degeneracy is also a measure of the environment in the phase space. In a strict sense, an energy eigenstate for the subsystem no longer exists (unless it is totally separated in terms of all possible interactions), so its energy E_1 will fail to show a sharp value in observations. Again, this effect is both classical and quantum, lying in the way the observed system is tested physically.

Having a distribution of possible energies, or in the quantum description, oscilla-tions between possible imagined reference eigenstates of the subsystem, we find the starting point of the ensemble view. We look for the conditional probability $P(E_1|E)$ in the ideal case described by a certain rational expression of phase space volume factors. The microcanonical definition of a fixed energy shell in N-dimensional phase space, viz.,

$$\Omega_N(E) \equiv \int d\Gamma_N \, \delta \, (E - H_N),   \tag{5.1}$$

suggests the following convolution property:

$$\Omega_{12}(E) = \int_0^E dE_1 \, \Omega_1(E_1) \, \Omega_2(E - E_1).   \tag{5.2}$$

This relation allows for the following normalized definition of the conditional prob-ability we seek:

$$P(E_1|E) \equiv \frac{\Omega_1(E_1)\Omega_2(E_2)}{\Omega_{12}(E)}.   \tag{5.3}$$

In view of (5.2), this is normalized to 1 with respect to integration over E_1, for an additive energy $E = E_1 + E_2$. For ample and frequent interactions with the envi-ronment, we could take $E = E_1 + E_2 + E_{int}$, although there is no common sense agreement on how E_{int} should be shared between the systems, i.e., on what exactly E_1 and E_2 should be. The very root of the thermal concept is to partition into statis-tically independent systems, and for such systems the product form (5.3) is adopted. It is delightful to observe that if, following Boltzmann, we define the entropy via the phase space volume by

$$S(E) \equiv \ln \frac{\Omega(E)}{\Omega_0},   \tag{5.4}$$

in units such that $k_B = 1$, with $\Omega_0 = (2\pi\hbar)^N$ as the reference phase space cell volume, then the conditional probability we are discussing is the exponential of the mutual information, i.e., the information one gains from a system by observing its complement in the whole. Expressing this in terms of the energy share $x_1 = E_1/E$, we have

$$P(E_1|E) = \frac{\Omega_1(x_1 E)\Omega_2((1-x_1)E)}{\Omega_{12}(E)} = e^{I(x_1;E)} , \qquad (5.5)$$

with

$$I(x_1; E) \equiv S_1(x_1 E) + S_2((1-x_1)E) - S(E). \qquad (5.6)$$

In the 'thermodynamic limit', one considers only the maximum of this conditional probability, singling out the subsystem energy $E_1 = x_1^{\max} E$. This delivers the equality of the energy derivatives of the entropies at this point, defining thermal equilibrium. For the same equation of state for the subsystem and reservoir, we obviously have $x_1^{\max} = 1/2$, corresponding to *equipartition*. A Gaussian approximation can sometimes be elaborated near this maximum point and used to estimate the leading order thermal fluctuations around the equilibrium parameters. However, for some special functions $S(E)$, and in particular for ideal gases defined by their kinetic motion alone, the full conditional distribution can be obtained exactly.

The thermodynamic limit includes the limit of an infinite reservoir, $E \to \infty$, and one considers

$$P(E_1|\infty) = \Omega_1(E_1)\rho(E_1) , \qquad (5.7)$$

where $\Omega_1(E_1)$ is the phase space volume for the observed subsystem and

$$\rho(E_1) = \lim_{E_1 \ll E} \frac{\Omega_2(E - E_1)}{\Omega_{12}(E)} \qquad (5.8)$$

is the thermal weight factor, interpreted as an occurrence frequency in a thermal ensemble of like systems. The following normalization holds:

$$\mathrm{Tr}\hat{\rho}_1 = \int_0^\infty \rho(E_1)\Omega_1(E_1)\,dE_1 = 1. \qquad (5.9)$$

Here the phase space measure differential of the subsystem is written in the form $d\Gamma_1 = \rho(E_1)dE_1$, where it is expressed in terms of the energy. We may learn by expressing this thermal weight factor in terms of entropies, too. An expansion in $E_1 \ll E$ delivers

$$\rho(E_1) = \lim_{E \to \infty} \exp\left[S_2(E) - E_1 S_2'(E) + \frac{1}{2}E_1^2 S_2''(E) + \cdots - S_{12}(E) \right]. \qquad (5.10)$$

In extensive systems, we usually have $S(E) \propto \mathscr{O}(E)$, and therefore $S'(E) \propto \mathscr{O}(1)$, $S''(E) \propto \mathscr{O}(1/E)$, and so on.[1] This leading order behavior gives rise to the following interpretation of the parameters in the well-known formula

$$\rho(E_1) = \frac{1}{Z} e^{-\beta E_1} . \tag{5.11}$$

Here the partition function, Z, is calculated as

$$Z = e^{S_{12}(\infty) - S_2(\infty)} = e^{S_1(\infty)} = \int_0^\infty \Omega(E_1) e^{-\beta E_1} \, dE_1. \tag{5.12}$$

In the limit of infinite energy in the total system, we have formally $\beta = S_2'(\infty)$ and $Z = e^{S_1(\infty)}$. In this way everything is expressed through characteristics of the subsystem, *except for the temperature, which is set by the environment*.

Several deviations from this classical result occur if the system size is kept finite. In order to demonstrate this, we discuss an analytically tractable case. For kinetic models, when the occupied phase space volume is determined by additive kinetic energies alone, ultrarelativistic degrees of freedom are described solely by their energy. Then in N phase space dimensions, the ansatz

$$\Omega_N(E) = f(N) E^{N-1} \tag{5.13}$$

is a good starting point. In order to satisfy the convolution property (5.2), we have

$$f(N) E^{N-1} = \int_0^E f(N_1) E_1^{N_1-1} f(N_2) E_2^{N_2-1} \, dE_1. \tag{5.14}$$

Since this power ansatz scales, it is natural to use the energy share variable $x = E_1/E$ in the integral. We arrive at

$$f(N) E^{N-1} = E^{(N_1-1)+(N_2-1)+1} f(N_1) f(N_2) \int_0^1 x^{N_1-1} (1-x)^{N_2-1} dx. \tag{5.15}$$

Clearly, the additive choice $N = N_1 + N_2$ eliminates the E-dependence, and the prefactor satisfies

$$f(N_1 + N_2) = f(N_1) f(N_2) B(N_1, N_2) , \tag{5.16}$$

---

[1] A notable exception is provided by black hole horizons, where $S(E) \sim E^2$, $S'(E) \sim E$, and $S''(E) \propto \mathscr{O}(1) > 0$, with a seemingly negative specific heat. For classical ideal gases, on the other hand, $S(E, N) = aN \ln(E/NT_0)$ and $\partial S/\partial E = aN/E = 1/T$, while $\partial^2 S/\partial E^2 = -aN/E^2 = -1/N c_V T^2$ delivers a positive and constant specific heat at fixed volumes: $c_N = 1/a$.

with Euler's beta-integral

$$B(N_1, N_2) = \int_0^1 x^{N_1-1}(1-x)^{N_2-1}dx = \frac{\Gamma(N_1)\Gamma(N_2)}{\Gamma(N_1+N_2)}. \tag{5.17}$$

The solution of (5.16) delivers $f(N) \propto 1/\Gamma(N)$, and finally

$$\Omega_N(E) = \frac{\xi^N}{\Gamma(N)}E^{N-1}. \tag{5.18}$$

The lessons from this study are several. The distribution of the subsystem energy is represented by an Euler beta distribution. The degrees of freedom N_1, on the other hand, follow a Bernoulli distribution for a fixed total of $N = N_1 + N_2$. Changing this condition to that of a fixed environment, N_2 is constant, and $P(E_1|E_2)$ converts into a negative binomial distribution in N_1 [4–6]:

$$P(E_1, N_1|E_2, N_2)dE_1 = \frac{\Gamma(N_1+N_2)}{\Gamma(N_1)\Gamma(N_2)} \frac{E_1^{N_1-1}E_2^{N_2-1}}{(E_1+E_2)^{N_1+N_2-1}} dE_1. \tag{5.19}$$

In particular, if both the subsystem and its counterpart reservoir were in energy eigenstates E_1 and E_2, respectively, and if further the reservoir dimensionality (degrees of freedom) N_2 were fixed, then the dimensionality of the subsystem, N_1, would fluctuate according to a negative binomial distribution. This situation can be typical when producing new particles in a thermal environment, where their number changes from event to event [7–11].

It is especially enlightening to consider the conditional probability for the single particle energy when the energy of the rest of the system is fixed:

$$P(E_1, 1|E_2, N-1) = \frac{N-1}{E_2}\left(1 + \frac{E_1}{E_2}\right)^{-(N-1)}. \tag{5.20}$$

This is a Tsallis–Pareto distribution in E_1, already suggested by Rolf Hagedorn to describe high energy particle spectra by putting $E_2 = (N-1)T$, $E_1 = |p_T|$, and $v = (N-1)$, which is not necessarily an integer. In these terms, the above formula looks like [12–21]

$$P(p_T, 1|vT, v) = \frac{1}{T}\left(1 + \frac{1}{v}\frac{|p_T|}{T}\right)^{-v}. \tag{5.21}$$

Finally, let us consider the limit $N \gg 1$, i.e., the limit $N_1 \ll N_2$, of the above. Now fixing $E_2/(N_2-1) = T$ with both E_2 and N_2 large, the conditional probability of finding a total energy E_1 for the N_1 degrees of freedom in the subsystem is

$$P(E_1, N_1|T \cdot \infty, \infty) = \frac{1}{T} \frac{(E_1/T)^{N_1-1}}{(N_1-1)!} e^{-E_1/T} . \tag{5.22}$$

On the one hand, this is an N_1-fold convolution of the simple exponential Boltzmann factor for the phase space with $N_1 - 1$ degrees of freedom, and on the other hand, a Poisson distribution in $N_1 - 1$. Indeed, with this probability density function, we obtain $\langle E_1 \rangle = N_1 T$ and $\Delta E_1^2/\langle E_1 \rangle = T$.

To end this section, we note that the ratio $\beta = N_1/E_1$ also has a distribution. In the general case, it follows a beta distribution, in the large reservoir limit a gamma distribution, and only in the case when the subsystem is also large will its variance really vanish, giving room for a sharp Dirac-delta distribution around the canonical parameter $\beta_{max} = 1/T$. The kinetic temperature we extract from measurements on single particle energy distributions (spectra) is just the expectation value of a fluctuating quantity.

5.2 Measurement and Randomization in Quantum Physics

Temperature is not given a priori in physics. In fact quantum systems, whose evolution is unitary, must recur in a state with a high overlap with their initial state after a long enough time (the *recurrence time*). Observation, identified in the Copenhagen school with projection onto an eigenstate, also takes a certain amount of time physically. Changing from a linear combination of several states to a given selected eigenstate of a particular operator, generally the Hamiltonian, must in principle be described by a unitary time evolution. As such, the initial state cannot ever be totally forgotten.

However, in quantum many-body systems with a large number of degrees of freedom, the state can most of the time be close to one described by a randomized phase density operator, such as the thermal state. The more degrees of freedom, the greater the possibility that the time necessary for forgetting the closeness of the initial state, the so-called *dephasing time*, will be much shorter than the *recurrence time*. Being close to a thermal density matrix is intended in the sense of a logarithmic-type entropic divergence functional. Exceptions are fully integrable systems, which periodically return to states arbitrarily close to the initial one. Other systems, in particular subsystems coupled to a much larger environment, practically never return too closely to initial states, and most time averages over times longer than the dephasing time behave as though they are close to a randomized state.

If the density matrix is reconstructed from measurements, each of which takes a certain time, say δt, then this itself has a diagonalizing effect. For simplicity, let us consider a general density matrix in terms of energy eigenstates at time t :

$$\hat{\rho}(t) = \sum_{i=0}^{\infty} C_{ij} e^{i(\omega_j - \omega_i)t} |i\rangle \langle j| . \tag{5.23}$$

It has the following general matrix elements:

$$\rho_{ij}(t) = e^{i(\omega_j - \omega_i)t} \rho_{ij}(0). \tag{5.24}$$

The energy eigenvalues are $E_i = \hbar\omega_i$ in this case. Since the time evolution is unitary, the relative magnitude of diagonal and off-diagonal density matrix elements cannot be changed by this in the long term, and only oscillations occur in the magnitudes of the matrix elements. On the other hand, time averaging results in an integral of the phase factors. Based on

$$\langle \rho_{ij} \rangle_{\delta t} \equiv \frac{1}{\delta t} \int_{-\delta t/2}^{+\delta t/2} \rho_{ij}(t) \, dt , \tag{5.25}$$

we obtain

$$\langle \rho_{ij} \rangle_{\delta t} = \frac{\sin\left(\dfrac{\omega_j - \omega_i}{2} \delta t\right)}{\dfrac{\omega_j - \omega_i}{2} \delta t} \rho_{ij}(0). \tag{5.26}$$

From this result it is obvious that, for sufficiently long averaging times, the off-diagonal elements of the initial density matrix become suppressed. Practically speaking, given the behaviour of the function $\sin(x)/x$, the characteristic time is

$$\delta t_{\text{diag}} = \frac{2\pi}{|\omega_i - \omega_j|} = \frac{h}{|E_i - E_j|}. \tag{5.27}$$

For a mixture of states with close energy eigenvalues, the necessary averaging time nevertheless diverges. The diagonal elements remain unchanged under time averaging:

$$\langle \rho_{ii} \rangle_{\delta t} = \rho_{ii}(0). \tag{5.28}$$

We conclude that time integral averaging cannot be achieved by unitary operations.

A physical system in the real world can seldom be separated from a complex environment. Theoretically, this means that the total Hamiltonian is not exactly known, or its eigenstates remain unknown in analytical form. The standard approach then considers a simple, solvable part of the Hamiltonian with its known eigenstates, and assumes that interactions with the environment (including the measuring apparatus) can be treated as small perturbations. A perturbative solution of the time-dependent Schrödinger equation in the energy eigenstate representation demonstrates this:

$$\left(H_{\text{solvable}} + H_{\text{pert}} - i\hbar \frac{\partial}{\partial t} \right) \sum_n C_n(t) \, e^{-it E_n/\hbar} \, |n\rangle = 0 \tag{5.29}$$

splits the Hamiltonian and uses the oscillating factors of the H_{solvable} part explicitly. Considering the Hilbert scalar product of this equation with the eigenstate $\langle k|$ of the unperturbed Hamiltonian, we derive a system of equations for the still time-dependent (perturbed) coefficients

$$i\hbar \frac{dC_k(t)}{dt} = \sum_n \langle k| H_{\text{pert}} |n\rangle \, C_n(t) \, e^{it(E_k - E_n)/\hbar} . \tag{5.30}$$

Here a modification of the basic oscillatory behavior of the mixing coefficients, $C_n(t)$, is caused by the interaction with the perturbing agents, describing transitions between the original energy eigenstates. Note, furthermore, that a complex oscillating factor also occurs here with the relative frequency $\omega = (E_k - E_n)/\hbar$.

Traditionally, the following case is considered: the quantum system has been unperturbed until the time $t = 0$ and then a weak and time-independent disturbance is switched on. In this case, one counts with $C_n(0) = \delta_{n,o}$, starting with a well-prepared unperturbed energy eigenstate $|o\rangle$. Then the solution for the time development of a small admixture of another eigenstate is described by the integral of (5.30) and is given by

$$C_k(t) = \frac{\langle k| H_{\text{pert}} |o\rangle}{i\hbar} \, e^{i\omega t/2} \, \frac{\sin(\omega t/2)}{\omega/2}. \tag{5.31}$$

From now on, $\omega = (E_k - E_o)/\hbar$. The probability of being in state $|k\rangle$ oscillates in time:

$$|C_k(t)|^2 = \frac{4}{(\hbar\omega)^2} \, |\langle k| H_{\text{pert}} |o\rangle|^2 \, \sin^2(\omega t/2). \tag{5.32}$$

The transition rate, taken as the time rate of change of the probability of being in the original state $|o\rangle$, is given by

$$\Gamma_{o \to k} = \frac{d}{dt} |C_k(t)|^2 = \frac{2}{\hbar^2} \, |\langle k| H_{\text{pert}} |o\rangle|^2 \, \frac{\sin \omega t}{\omega}. \tag{5.33}$$

This result can be viewed in several ways. For short times ($t \ll \pi/\omega$, describing the immediate reaction to the perturbation), the decay rate of the original state $|o\rangle$ is proportional to the time. When it grows too large, the assumption that the perturbation is small can no longer be upheld. Based on (5.32), around this time, the probability of finding the system in the new state, where the perturbation tends to push it, is maximal. On the other hand, for closely spaced energy eigenvalues (dense spectrum), $\omega \to 0$ is the relevant approximation, and $\sin(\omega t)/\omega \to t$ holds once again.

Considering now a continuum of states with the density $S(\omega)$ per energy interval, the total transition rate from an initial state $|i\rangle$ to a final state $|f\rangle$ is given by the integral

$$\Gamma_{i \to f} = \frac{2}{\hbar^2} \int\limits_{-\infty}^{+\infty} d\omega \, S(\omega) \, |\langle f| H_{\text{pert}} |i\rangle|^2 \, \frac{\sin \omega t}{\omega}. \tag{5.34}$$

One usually considers the following approximation. The new integration variable is $x = \omega t$. In the limit $t \to \infty$, the x-integral over the finite range $[-\pi/2, +\pi/2]$ mimics the infinite range ω-integral[2]. We thus arrive at the approximation

$$\int_{-\infty}^{+\infty} S(\omega) \frac{\sin \omega t}{\omega} \, d\omega \approx \int_{-\pi/2}^{+\pi/2} S(x/t) \frac{\sin x}{x} \, dx \approx S(0) \frac{\pi}{2}. \tag{5.35}$$

Using this result, the main contribution under the integral in (5.34) becomes:

$$\Gamma_{i \to f} \approx \frac{\pi}{\hbar^2} S(0) \left| \langle f | H_{\text{pert}} | i \rangle \right|^2. \tag{5.36}$$

It is a general assumption that the spectral function of the frequency difference is slowly changing in comparison to the fast oscillations stemming from the energy eigenvalue distance between the states. The constant decay rate in the above approximation results in an exponential decay law in time for particles, and is known as Fermi's golden rule. Here only the magnitude of the transition matrix element counts [22].

Time averages or integrals, similar to the 'projections' by a measurement, enhance certain diagonal elements of a density matrix at the expense of others. Interactions with the environment seem to happen with a random phase, leading to a thermal-type density matrix after a while. So why worry about recurrence? In quantum physics, the question concerns the recurrence of the quantum phase. The evolution operator, given as a general solution to the Schrödinger equation, is unitary. Given two arbitrary quantum states as vectors in the Hilbert space, we construct a unitary operator mapping one vector to the other. This operator has a spectrum representable on the perimeter of a circle, while its eigenvectors form a complete orthogonal system. A unitary operator on the other hand is always interpretable as a complex exponential of the form $U = \exp(it H/\hbar)$, involving a given time duration t times a corresponding frequency $\omega = E/\hbar$. Whenever the ratio of two such frequencies is irrational, all states will be visited in due of time. Most randomly picked Hamiltonians result in such a system. In the null-measure set of rational frequency ratios, there is a further conserved quantity selecting a lower dimensional subset of the Hilbert space.

In the following, we carry out the construction of a unitary transformation between two vectors of the Hilbert space with the same norm. In order to proceed, we seek the unitary evolution operator between two given states both normalized to 1. Let these states both be a linear combination of two orthogonal states:

$$|z_1\rangle = A |0\rangle + B |1\rangle ,$$
$$|z_2\rangle = C |0\rangle + D |1\rangle . \tag{5.37}$$

The states $|0\rangle$ and $|1\rangle$ form an orthonormal basis.

---

[2] The choice of this range is motivated by the fact that the function $\sin x / x$ differs from zero mainly in this interval.

We define four operators acting on the $|0\rangle$, $|1\rangle$ basis in such a way as to form a C_2 (quaternion) algebra:

$$P \equiv |1\rangle \langle 1| + |0\rangle \langle 0| \,,$$
$$V \equiv |1\rangle \langle 0| - |0\rangle \langle 1| \,,$$
$$W \equiv i\,|1\rangle \langle 0| + i\,|0\rangle \langle 1| \,,$$
$$U \equiv i\,|1\rangle \langle 1| - i\,|0\rangle \langle 0| \,. \tag{5.38}$$

These operators satisfy the following relations:

$$P^2 = P \,, \qquad V^2 = -P \,, \qquad W^2 = -P \,, \qquad U^2 = -P \,,$$
$$PV = VP = V \,, \qquad PW = WP = W \,, \qquad PU = UP = U \,, \tag{5.39}$$
$$U = VW = -WV \,, \qquad W = UV = -VU \,, \qquad V = WU = -UW.$$

The operators V, W, and U are anti-Hermitian, while P is Hermitian. Indeed, expressed as matrices in the above basis, we have $P = 1$, $V = i\sigma^2$, $W = i\sigma^1$, $U = i\sigma^3$, where σ_i, $i = 1, 2, 3$ are the Pauli matrices.

Here P and U are composed of projectors, while V and W utilize the step operators. Regarding the $|0\rangle$, $|1\rangle$ basis as a physical quantum system with two orthogonal states, we define the step-down operator as

$$a \equiv |0\rangle \langle 1| \,, \qquad a^\dagger = |1\rangle \langle 0| \,, \tag{5.40}$$

satisfying $a\,|1\rangle = |0\rangle$, $a\,|0\rangle = 0$ and $a^\dagger\,|1\rangle = 0$, $a^\dagger\,|0\rangle = |1\rangle$, and note that $V = a^\dagger - a$, $W = i(a^\dagger + a)$.

A unitary matrix is generally constructed as an exponential of $i\omega \mathbf{n}\boldsymbol{\sigma}$. In our special case, we seek a Glauber-type construction, so we set the coefficient of U in the exponent equal to zero. Then,

$$e^{itH} = e^{\omega\left[\sin\Theta\,(\cos\varphi\,V + \sin\varphi\,W) + \cos\Theta\,U\right]} \tag{5.41}$$

simplifies the Hamiltonian to

$$itH = \omega\,(\cos\varphi\,V + \sin\varphi\,W) = \omega\,e^{i\varphi}\,a^\dagger - \omega\,e^{-i\varphi}\,a. \tag{5.42}$$

This is a Glauber shift operator with the complex coefficient $z = \omega\,e^{i\varphi}$.

Using the well-known result regarding the su(2) algebra, we have

$$e^{itH} = e^{za^\dagger - z^*a} = \cos\omega\,P + \frac{\sin\omega}{\omega}\,\left(za^\dagger - z^*a\right). \tag{5.43}$$

Given the actions of the step operators on the basis states, we easily obtain

$$e^{itH} |0\rangle = \cos\omega \, |0\rangle + \sin\omega \, e^{i\varphi} \, |1\rangle \; ,$$
$$e^{itH} |1\rangle = \cos\omega \, |1\rangle - \sin\omega \, e^{-i\varphi} \, |0\rangle \; . \tag{5.44}$$

Now we obtain the complex coefficients A, B, C, and D. Normalization of the $|z_1\rangle$ and $|z_2\rangle$ states requires $|A|^2 + |B|^2 = 1$ and $|C|^2 + |D|^2 = 1$. However, the overlap between these two states is a given complex number (with magnitude less than or equal to one):

$$\langle z_2| \, z_1\rangle = AC^* + BD^* . \tag{5.45}$$

Requiring now that the above evolution send $|z_1\rangle$ exactly to $|z_2\rangle$, we have

$$e^{itH} |z_1\rangle = A\left(\cos\omega \, |0\rangle + \sin\omega \, e^{i\varphi} \, |1\rangle\right) + B\left(\cos\omega \, |1\rangle - \sin\omega \, e^{-i\varphi} \, |0\rangle\right)$$
$$= C \, |0\rangle + D \, |1\rangle = |z_2\rangle \; , \tag{5.46}$$

and we obtain the coefficients C and D in terms of A and B as follows:

$$C = A\cos\omega - B\sin\omega \, e^{-i\varphi} \; ,$$
$$D = B\cos\omega + A\sin\omega \, e^{i\varphi} \; . \tag{5.47}$$

The real and imaginary parts of the state overlap are then expressed as

$$\langle z_2| \, z_1\rangle = AC^* + BD^* = \cos\omega - 2i\sin\omega \, \Im \left(AB^* \, e^{i\varphi}\right) . \tag{5.48}$$

As a matter of fact, the simple choice $A = 1/\sqrt{2}$ and $B = -1/\sqrt{2}$ is a sufficient basis for the calculations. With this choice.

$$\Re \, \langle z_2| \, z_1\rangle = \cos\omega \; ,$$
$$\Im \, \langle z_2| \, z_1\rangle = \sin\omega \sin\varphi \; . \tag{5.49}$$

From this and the content of the previous section, we see that a non-unitary density operator can only describe a subsystem of a bigger one, with possible energy exchange between the two. In thermal field theory, we are especially interested in some simplified assumptions about this connection, when the influence of the environment can be attributed to a random noise.

5.3 Additive and Multiplicative Noise

The behavior of a subsystem under noise, under random influences from its environment, is the description on a level between the microscopic quantum dynamics and the macroscopic thermal behavior. At this medium level it contributes to the understanding of the emergence of thermal effects. The absolute temperature treated as an average kinetic energy per degree of freedom in such models arises as a given

property of the noise, and in the long term, time-averages turn out to be governed by
ensemble averages set by the environment. Before turning to field theoretical com-
putation techniques accounting for such noise effects, let us give a general overview
of the actors and terms in this piece of drama.

Physical systems with many degrees of freedom are often observed only indi-
rectly, through some selected, 'slow' components. The ignored details of the time
evolution of the rest will then be recognized in some average, 'effective' parameters.
Noise develops in time unpredictably. It is a stochastic component, described by
the expectation values of its amplitude and some of its correlation characteristics.
Randomness and noise are synonyms. The first emphasizes the mathematical nature,
the latter the physical properties. The simplest way to introduce noise effects in a
deterministic description of dynamics is to add them as source terms in the equations
of motion. This then also affects the variational principle behind these equations of
motion. Noise equations exploit a separation of slow and fast variables, although
the exact borderline is not easily defined. Using a weak coupling philosophy, when
the back-reaction of the observed slow subsystem on the noise can be neglected,
akin to considering infinite reservoirs as part of the thermodynamical limit, it is an
acceptable approximation to keep the noise properties (autocorrelation parameters)
constant in time.

The basic idea behind the Feynman–Vernon path integral [23], treating quantum
systems under the influence of noise, is a distinction between slow variables, Φ fields,
and the fast-changing background ones, η. The action then contains three types of
contributions: one depending only on the observed fields, one due to their coupling
to the rest, and one purely described by the background:

$$S[\Phi, \eta] = S_{slow}[\Phi] + S_{coupled}[\Phi, \eta] + S_{background}[\eta] . \qquad (5.50)$$

The simplest version of such an approach delivers an additive noise by assuming

$$S_{coupled}[\Phi, \eta] = g \int \Phi \eta \, dt . \qquad (5.51)$$

It gives rise to a modified field equation for the slow variable just by adding a simple
source term:

$$\frac{\delta S_{slow}}{\delta \Phi} = -g\eta . \qquad (5.52)$$

For higher order couplings, like, e.g., $\eta V(\Phi)$, the noise effect may also occur in
terms containing the Φ field itself in an expression; these are known generically as
multiplicative noise approaches. The Feynman path integral containing all degrees
of freedom is then converted into a form featuring the evolution of the slow variables
alone, with an effective action:

$$G = \int \mathscr{D}\Phi \left[\int \mathscr{D}\eta \, e^{\frac{i}{\hbar} S[\Phi, \eta]} \right] = \int \mathscr{D}\Phi \, e^{\frac{i}{\hbar} S_{eff}[\Phi]} . \qquad (5.53)$$

An underlying assumption for making the separation of noise practicable is that their evolution should be fast on the time scale of the evolution of the observed field, Φ. Moreover they must also quickly forget their previous states, i.e., their correlations must decay much faster than the time scale of the observation. Then, formally, the η degrees of freedom are included only via their distribution (Wigner functional)

$$P[\eta] = e^{\frac{i}{\hbar} S_{\text{background}}[\eta]} .$$ (5.54)

The simplest noise is white noise, when this distribution is regarded as Gaussian at any instant of time, independently of any previous times (this being the Markovian property of the noise). In this case, the limiting Wigner distribution is positive:

$$P[\eta] = N e^{-\int \eta^2(t)/2\sigma^2 dt} ,$$ (5.55)

with an unspecified normalization factor. In this case, the assumed background action is purely imaginary: according to the philosophy of Schrödinger's variational principle, it is a classically impossible possibility.

Gaussian dependencies on background fields can be path-integrated exactly. With the above assumption, the effective (noisy) action for the extracted slow variables is then given formally by

$$S_{\text{eff}}[\Phi] = S_{\text{slow}}[\Phi] + \frac{\hbar}{i} \ln \int \mathscr{D}\eta \, P[\eta] \, e^{\frac{i}{\hbar} S_{\text{coupled}}[\Phi,\eta]} .$$ (5.56)

For Gaussian additive noise, meaning a linear coupling to the slow variables, the path integral part under the logarithm can be formally evaluated. How does it change the original part of the action corresponding to the slow variables alone? A simple demonstration can be given by inspecting the characteristic function of the Gauss distribution, the expectation value of $\langle e^{bz} \rangle$ for a variable z distributed according to

$$P(z) = \frac{1}{\sqrt{2\pi \langle z^2 \rangle}} e^{-z^2/2\langle z^2 \rangle} .$$ (5.57)

One easily obtains

$$\langle z^{2k+1} \rangle = 0 , \qquad \langle z^{2k} \rangle = (2k-1)!! \langle z^2 \rangle^k ,$$ (5.58)

for k a positive integer and $(2k-1)!!$ the product of odd numbers alone. The characteristic function can be calculated as follows:

$$\langle e^{bz} \rangle = \sum_{n=0}^{\infty} \frac{b^n}{n!} \langle z^n \rangle = \sum_{k=0}^{\infty} \frac{(2k-1)!!}{(2k)!} b^{2k} \langle z^2 \rangle^k = e^{b^2 \langle z^2 \rangle/2} .$$ (5.59)

A remarkable consequence of this result is its application to the Taylor expansion of a function with shifted variable [24–27]:

$$\langle\, f(a+z)\,\rangle = \langle\, e^{z\frac{\partial}{\partial a}}\,\rangle f(a) = e^{\frac{1}{2}\langle z^2\rangle\frac{\partial^2}{\partial a^2}}\, f(a)\,. \tag{5.60}$$

The effective action of slow fields under the influence of Gaussian noise can then be calculated for a linear coupling with an analogous expression using functional derivatives:

$$e^{\frac{i}{\hbar}S_{\text{eff}}} = \langle\, e^{\frac{i}{\hbar}(S[\varPhi]-g\int\varPhi\eta)}\,\rangle = e^{-\frac{g^2}{2\hbar^2}\int dt_1 dt_2\,\langle\,\eta(t_1)\eta(t_2)\,\rangle\frac{\delta}{\delta\varPhi(t_1)}\frac{\delta}{\delta\varPhi(t_2)}}\, e^{\frac{i}{\hbar}S[\varPhi]}\,. \tag{5.61}$$

In fact, this approach considers the generating functional,

$$Z[J] \equiv \int \mathscr{D}\varPhi\, e^{\frac{i}{\hbar}(S[\varPhi]-\int J\varPhi)}\,, \tag{5.62}$$

with an added noise

$$\langle\, Z[J+\eta]\,\rangle = \langle\, e^{\int\eta\frac{\partial}{\partial J}}\,\rangle Z[J] = e^{\frac{1}{2}\int N_{12}\frac{\partial}{\partial J_1}\frac{\partial}{\partial J_2}}\, Z[J] \tag{5.63}$$

and the noise propagator

$$N_{12} = \langle\, \mathscr{T}\eta(x_1)\eta(x_2)\,\rangle\,, \tag{5.64}$$

with various time-ordering prescriptions in the Keldysh formalism discussed previously. Here we have suppressed the details of various integrals over the spacetime coordinates in our notation.

5.4   Oscillator Bath

The simplest known example for the behavior of a slow, heavy particle in a wildly fluctuating background is Brownian motion [28]. The description here can be given either by a Langevin equation [29] with an additive noise term and an immediate damping force proportional to the velocity, or equivalently in a corresponding Fokker–Planck equation with diffusion and drift terms [30, 31]. There is a plethora of excellent textbooks devoted to the details of these descriptions, so no need to repeat them here [32]. Instead, we present a description of slow degrees of freedom coupled to a (large) number of oscillators. The calculations here are somewhat more involved, but from the field theory point of view, the approach comes closer to those used in this book.

We consider a classical Hamiltonian description of the coupled dynamics; this will give a good idea of the results expected in the semiclassical approximation to field theory. The analysis is based on [33, 34]. Let the Hamiltonian of the slow degree of freedom be $H_s(P, Q)$, where Q is the corresponding canonical coordinate and P the associated momentum. It will not be further specified here. We investigate

$$H = H_s(P, Q) + \sum_i \left(\frac{1}{2m_i} p_i^2 + \frac{m_i \omega_i^2}{2} q_i^2 \right) - Q \sum_i g_i q_i . \tag{5.65}$$

The equations of motion are

$$\dot{Q} = \frac{\partial H_s}{\partial P} , \qquad \dot{P} = -\frac{\partial H_s}{\partial Q} + \sum_i g_i q_i ,$$
$$\dot{q}_i = \frac{p_i}{m_i} , \qquad \dot{p}_i = -m_i \omega_i^2 q_i + g_i Q . \tag{5.66}$$

For this simple coupling, the equation of motion for the individual oscillators is analytically solvable. It is in the form of a driven oscillator equation,

$$\ddot{q}_i + \omega_i^2 q_i = \frac{g_i}{m_i} Q(t) . \tag{5.67}$$

Its solution contains a homogeneous part, describing the oscillations without the external drive term, and an integral form containing the retarded Green's function (propagator) of the oscillator:

$$q_i(t) = q_i(0) \cos \omega_i t + \frac{p_i(0)}{m_i \omega_i} \sin \omega_i t + \frac{g_i}{m_i \omega_i} \int_0^t dt' \sin \omega_i (t - t') Q(t') . \tag{5.68}$$

For later interpretation of the various terms, let us rewrite here the main factor in the inhomogeneous solution using partial integration:

$$\omega_i \int_0^t dt' \, Q(t') \sin \omega_i (t - t') = Q(t) - Q(0) \cos \omega_i t - \int_0^t dt' \dot{Q}(t') \cos \omega_i (t - t') . \tag{5.69}$$

In this form, the integral involves $\dot{Q}$, making it a linear functional of the velocity of the slow degree of freedom. Using this, the source term for $\dot{P}$ in (5.66), viz., $\eta \equiv \sum_i g_i q_i(t)$, becomes

$$\eta = \sum_i g_i q_i^{\text{hom}}(t) + \sum_i \frac{g_i^2}{m_i \omega_i^2} [Q(t) - Q(0) \cos \omega_i t] - \int_0^t dt' 2\Gamma(t - t') \dot{Q}(t') . \tag{5.70}$$

Here we have introduced the damping kernel

$$\Gamma(t - t') \equiv \sum_i \frac{g_i^2}{2m_i\omega_i^2} \cos \omega_i (t - t') \,. \tag{5.71}$$

With all these preparations, we are now ready to turn to the equation of motion for the slow variable. It features a modification of the slow Hamiltonian, a dissipation force integral, and a rest term, which we shall call 'noise'. The exact equation of motion for the slow degree of freedom (often called a pointer) is then

$$\dot{P} + \frac{\partial H_s}{\partial Q} = \alpha Q - 2 \int_0^t dt' \Gamma(t - t') \dot{Q}(t') + \xi(t) \,. \tag{5.72}$$

We have introduced $\alpha = \sum_i g_i^2/m_i\omega_i^2$ as a shorthand notation and note that the linear term αQ on the right-hand side of the above equation only modifies the slow Hamiltonian to[3]

$$H_{\mathrm{mod}}(P, Q) = H_s(P, Q) - \frac{1}{2}\alpha Q^2 \,. \tag{5.73}$$

In order to make physical sense of this, we have to assume that the couplings are weak enough not to destroy the positivity of this Hamiltonian. The damping kernel $\Gamma(t - t')$ has already been defined above in (5.71). Finally, we are left with the rest term

$$\xi(t) \equiv \sum_i \left[\left(g_i q_i(0) - \frac{g_i^2}{m_i\omega_i^2} Q(0) \right) \cos \omega_i t + \frac{g_i p_i(0)}{m_i\omega_i} \sin \omega_i t \right] \,. \tag{5.74}$$

Its general form is just a sum over formal oscillator amplitudes

$$\xi(t) = \sum_i A_i \cos(\omega_i t - \delta_i) \,. \tag{5.75}$$

Without compromising the general nature of our analysis, we assume $Q(0) = 0$ from now on. This simplifies the formulas without losing any essential physics. In this case, the amplitude squared is just related to the oscillator energy:

$$A_i^2 = g_i^2 \left[q_i^2(0) + \frac{p_i^2(0)}{m_i^2\omega_i^2} \right] = \frac{2g_i^2}{m_i\omega_i^2} E_i(0) \,, \tag{5.76}$$

with the initial oscillator energies

$$E_i(0) = \frac{p_i^2(0)}{2m_i} + \frac{m_i\omega_i^2}{2}q_i^2(0) \,. \tag{5.77}$$

---

[3]It has a destabilizing effect for Hamiltonians $H_s(P, Q)$ with an energy minimum at $Q = 0$.

The relative phases are determined by $\tan \delta_i = p_i(0)/m_i \omega_i q_i(0)$. In most cases, the initial states of the bath oscillators are unknown. Only their average properties or a probability distribution of different energies in a thermal state are available as information. These determine the properties of the noise term.

We shall consider here two physical scenarios:

- In a thermal initial state of bath oscillators, both the $q_i(0)$ and the $p_i(0)$ are Gaussian distributed according to the probability factors $\exp\left[-E_i(0)/T\right]$.
- Considering uniformly random phases δ_i.

In both cases, $\langle \xi(t) \rangle = 0$, either due to $\langle q_i(0) \rangle = 0$ and $\langle p_i(0) \rangle = 0$ or due to $\langle \cos(\omega_i t - \delta_i) \rangle = 0$.

We are most interested in the correlation of the noise, viz.,

$$C(t, t') \equiv \langle \xi(t)\xi(t') \rangle . \tag{5.78}$$

Note that

$$\langle A_i A_j \cos(\omega_i t - \delta_i) \cos(\omega_j t' - \delta_j) \rangle = \frac{1}{2}\delta_{ij} A_i^2 \cos \omega_i (t - t') , \tag{5.79}$$

with nonzero result only for $i = j$, due to the identity

$$\int_{-\pi}^{+\pi} \frac{d\delta}{2\pi} \cos(a - \delta) \cos(b - \delta) = \frac{1}{2} \int_{-\pi}^{+\pi} \frac{d\delta}{2\pi}\left[\cos(a + b - 2\delta) + \cos(a - b)\right]$$

$$= \frac{1}{2} \cos(a - b) . \tag{5.80}$$

Bringing together the definition of the damping kernel, the relation between the amplitude and the individual oscillator energies, and this observation (5.79), the following fluctuation–dissipation relation holds in a random phase approximation:

$$C(t, t') = 2\langle E_i(0) \rangle \Gamma(t - t') . \tag{5.81}$$

For a thermal ensemble of oscillator initial states, one naturally has $\langle E_i(0) \rangle = T$, but this result is more general. The 'fluctuation' part, approximated by noise, gives energy to the slow motion degrees of freedom, while the 'dissipation' part takes it away. The final evolution converges to a stationary equilibrium between these two processes. A pointer, e.g., the momentum of the soft degree of freedom, follows an erratic path, but the average energy related to it has an expectation value. Moreover, such a pointer may follow a distribution which looks thermal in many cases.

This is also a mechanism by which temperature may emerge from microscopic motion. A simple analytic model behaving this way is described by the Langevin equation and the corresponding Fokker–Planck equation: a damping force, proportional to the velocity of the Brownian particle, is counterbalanced by white noise. Then,

$$\dot{p} + \Gamma \frac{p}{m} = \xi \ , \tag{5.82}$$

with $\langle \xi(t) \rangle = 0$, $\langle \xi(t)\xi(t') \rangle = 2D\delta(t - t')$, leads to the evolution of an ensemble of momenta, $f(p)$, governed by drift and diffusion terms:

$$\frac{\partial f}{\partial t} + \frac{\partial}{\partial p} \Gamma \frac{p}{m} f + \frac{\partial^2}{\partial p^2} Df = 0 \ , \tag{5.83}$$

which has the stationary solution

$$f(p, t \to \infty) \ \longrightarrow \ f(0)\, e^{-\Gamma p^2/2mD} \tag{5.84}$$

for p-independent Γ and D. Interpreting this as a distribution of the kinetic energy, $E = p^2/2m$, we realize that this is a Boltzmann–Gibbs distribution with temperature $T = D/\Gamma$. This 'Einstein temperature' is then constructed from the dissipation and diffusion constants. Here $D = T\Gamma$ is the fluctuation–dissipation relation.

In more complex cases, however, this is not the final answer. Since the dissipation and the noise both emerge from the coupling to the fast degrees of freedom, and this coupling can in general be anharmonic (nonlinear), it is not a surprise that the parameters of the environmental forces, i.e., Γ and D, may depend on the kinetic energy of the slow motion itself. This back-coupling to the noise makes the slow motion of the soft degree of freedom more interesting: the effect can be a non-exponential (non-Boltzmannian) distribution of the pointer particle's kinetic energy.

A slight generalization of the Langevin equation is as follows. We may consider several degrees of freedom, described by the momenta p_i (or field component variables). The dissipative force, in general a friction, is proportional to the velocity, at least in some leading order approximation. The general velocity is nothing else than the partial derivative of the total soft energy with respect to the corresponding momentum, i.e., $v_i = \partial E/\partial p_i$. From this, the Langevin equation has the form

$$\dot{p}_i + \Gamma_{ij}(E) \frac{\partial E}{\partial p_j} = \xi_i \ , \tag{5.85}$$

with a general correlation of the noise

$$\langle \xi_i(t)\xi_j(t') \rangle = 2D_{ij}(E)\delta(t - t') \ , \tag{5.86}$$

which may be energy dependent (this is a particular type of colored noise). The corresponding Fokker–Planck equation in this case has a stationary solution depending solely on the soft energy: $f(p_i, t \to \infty) = f(E)$. The generalized fluctuation–dissipation relation is given by [24, 35]

$$\Gamma_{ij}(E)f(E) + D_{ij}(E)f'(E) + D'_{ij}(E)f(E) = 0 \ . \tag{5.87}$$

For a single soft degree of freedom this equation can be solved formally to give

$$f(E) = \frac{D(0)}{D(E)} f(0) \exp \left[- \int_0^E \frac{\Gamma(U)}{D(U)} dU \right] . \tag{5.88}$$

For E-independent D and Γ, the Boltzmann distribution arises once again. However, considering a linear $D(E) = D_0 + D_1 E$ 'colored' noise, frequently arising as multiplicative noise in dynamical systems [36], one obtains a Tsallis–Pareto distribution of the soft energy:

$$f(E) = f(0) \left(1 + \frac{D_1}{D_0} E \right)^{-1-\Gamma/D_1} . \tag{5.89}$$

In the limit $D_1 \to 0$, of course, this formula goes over to the Boltzmann factor. Comparing with the original notation for the parameters,

$$f(E) = f(0) \left(1 + \frac{q-1}{T} E \right)^{-q/(q-1)} , \tag{5.90}$$

we may identify $T = D_0/\Gamma$ (as the Einstein temperature) and $q = 1 + D_1/\Gamma$ (as the Tsallis parameter). The fluctuation–dissipation relation can also be written in an elegant 'self-averaging' form:

$$D_{ij}(E) = \frac{1}{f(E)} \int_E^\infty \Gamma_{ij}(U) f(U) dU . \tag{5.91}$$

5.5 Emerging Noise in the Keldysh Formalism

With this powerful method in hand, we can take a look at different phenomena of the real world and try to explain them in the language of quantum field theory.

The primary phenomenon that we examine in this section is the presence of *noise*. In any measurement, the measured data are not 'precise', meaning that they differ from the theoretical values. But repeating the same experiment several times results in statistics, a distribution of the measured values where the mean value agrees with the expected value of the theory (provided the theory was correct). The fluctuation around the theoretical value can sometimes be visualized, as in the case of Brownian motion or in the shot noise of a CRT monitor [37, 38].

By attempting to give an account of this phenomenon, we realize that when explaining an observation with theory, we always make some simplifications. We concentrate on the most important (the most relevant) effects and assign the corresponding operator to them in the quantum theory. But there are countless (*de facto* infinitely many) irrelevant effects that are totally negligible taken separately, but, due to their large number, may have a cumulative effect. This can be taken into account using statistical methods.

To understand this in the language of path integrals, we consider a model where, besides the relevant φ field, there also appears an irrelevant, unmeasured χ field[4]. For simplicity, we assume a single interaction as in

$$\mathcal{L} = \frac{1}{2}\varphi(-\partial^2 - m^2)\varphi + \frac{1}{2}\chi(-\partial^2 - m^2)\chi + \frac{g}{2}\varphi\chi^2 . \tag{5.92}$$

In the real-time formalism, the complete Lagrangian is given by

$$\bar{\mathcal{L}} = \mathcal{L}(\varphi_1, \chi_1) - \mathcal{L}(\varphi_2, \chi_2) = \mathcal{L}\left(\varphi_r + \frac{\varphi_a}{2}, \chi_r + \frac{\chi_a}{2}\right) - \mathcal{L}\left(\varphi_r - \frac{\varphi_a}{2}, \chi_r - \frac{\chi_a}{2}\right) . \tag{5.93}$$

A crucial property of this fundamental Lagrangian is that it is antisymmetric under the interchange of contours $1 \leftrightarrow 2$, or, in the R/A formalism, it is odd in the advanced fields. In the present case we have, omitting total divergences:

$$\bar{\mathcal{L}} = \varphi_a(-\partial^2 - m^2)\varphi_r + \chi_a(-\partial^2 - m^2)\chi_r + \frac{g}{2}\varphi_a\chi_r^2 + g\varphi_r\chi_a\chi_r + \frac{g}{8}\varphi_a\chi_a^2 . \tag{5.94}$$

As already mentioned, the χ field cannot be measured, so we will compute only φ expectation values. Their generating functional reads

$$Z[J] = \int \mathcal{D}\varphi\mathcal{D}\chi \, e^{iS[\varphi,\chi] + \int J\varphi} . \tag{5.95}$$

In this formula the χ field can be integrated out, modifying the action. As a result we have

$$e^{iS_{\text{eff}}[\varphi]} = \int \mathcal{D}\chi \, e^{iS[\varphi,\chi]} = e^{iS[\varphi,\chi=0]} \int \mathcal{D}\chi \, e^{iS_0[\varphi,\chi]} e^{iS_{\text{int}}[\varphi,\chi]} . \tag{5.96}$$

Diagrammatically, we need to consider the connected diagrams where the internal lines are χ-propagators, and we have a φ background:

$$S_{\text{eff}} = S[\varphi, \chi = 0] - i\langle e^{iS_{\text{int}}[\varphi,\chi]} - 1 \rangle_{\text{connected}} . \tag{5.97}$$

The first nontrivial correction affects the quadratic part of the original action:

$$\delta S_2 = -\frac{i}{2}\langle (iS_{\text{int}}[\varphi,\chi])^2 \rangle_0 = g^2\varphi_a\delta K\varphi_r + \frac{ig^2}{2}\varphi_a D\varphi_a , \tag{5.98}$$

---

[4]By analogy with the previously discussed oscillator bath, $\varphi \sim Q$ and $\chi \sim q_i$.

where, in momentum space,

$$\delta K(p) = \frac{i}{2} \langle \chi_r^2 (\chi_r \chi_a) \rangle(p) ,$$

$$D(p) = \frac{1}{4} \langle \chi_r^2 \chi_r^2 \rangle(p) + \frac{1}{32} \langle \chi_r^2 \chi_a^2 \rangle(p) + \frac{1}{32} \langle \chi_a^2 \chi_r^2 \rangle(p) . \qquad (5.99)$$

In equilibrium, these are expressed by the equilibrium quantum distributions and the spectral functions of the fast field χ:

$$\delta K(p) = -\int \frac{d^4 k}{(2\pi)^4} G_{ra}(p-k) i G_{rr}(k) = -\int \frac{d^4 k}{(2\pi)^4} G_{ra}(p-k) \left[\frac{1}{2} + n(k_0) \right] S(k) ,$$
$$(5.100)$$

and

$$D(p) = \frac{1}{2} \int \frac{d^4 k}{(2\pi)^4} \left[i G_{rr}(p-k) i G_{rr}(k) + \frac{1}{4} S(p-k) S(k) \right]$$

$$= \frac{1}{2} \int \frac{d^4 k}{(2\pi)^4} \left[\frac{1}{2} [1 + n(k_0) + n(p_0 - k_0)] + n(k_0) n(p_0 - k_0) \right] S(k) S(p-k) .$$
$$(5.101)$$

To our surprise, the correction δS_2 of (5.98) is formally different from the quadratic part of the fundamental Lagrangian. First of all, there appears a term that is *even in* ϕ_a, and this term is purely imaginary, since $D(p)$ is real. Moreover, the correction to the kernel, viz., $\delta K(p)$, also contains an imaginary part. Because of these properties, the system in which a degree of freedom is integrated out cannot be regarded as a fundamental theory.

The latter problem, $\Im \delta K(p) \neq 0$, means that the retarded propagator in real time has some damping factor in it, so the time evolution is not unitary and the dynamics—at this level of approximation— reflects an open system. But how can we interpret the appearance of a term that is even in φ_a? Following Feynman and Vernon [23], we represent this term as an integral over a Gaussian-distributed Markovian background field, i.e., white noise:

$$\prod_p e^{-\frac{g^2}{2} \varphi_r^*(p) D(p) \varphi_r(p)} \sim \int \prod_p d\xi_p e^{-\frac{1}{2g^2 D(p)} \xi^*(p)\xi(p) - \frac{1}{2}\xi^*(p)\varphi_r(p) - \frac{1}{2}\xi(p)\varphi_r^*(p)} . \qquad (5.102)$$

Here $\xi(p)$ represents a *stochastic field variable* with Gaussian distribution function, i.e., white noise with correlation

$$\langle \xi^*(p)\xi(q) \rangle = g^2 D(p)\delta(p-q) . \qquad (5.103)$$

This representation therefore reads

$$e^{-\frac{g^2}{2}\int \frac{d^4 p}{(2\pi)^4} \varphi_r^*(p) D(p) \varphi_r(p)} = \langle e^{-\frac{1}{2}\int \frac{d^4 p}{(2\pi)^4} (\xi^*(p)\varphi_r(p)+\xi(p)\varphi_r^*(p))} \rangle_\xi . \tag{5.104}$$

So we conclude that, at least to quadratic order, the effective action coming from integrating over the χ degrees of freedom results in the appearance of a self-energy δK, which leads to a non-unitary time evolution, and in addition there is a stochastic force term, white noise, and *this makes up for the loss of unitarity*.

The damping and noise terms are intertwined. Using the form of the Bose–Einstein statistics

$$n(k_0) = \frac{1}{e^{\beta k_0} - 1} \quad \Longrightarrow \quad e^{\beta k_0} = 1 + \frac{1}{n(k_0)} , \tag{5.105}$$

and splitting $k_0 = p_0 + (k_0 - p_0)$, we find

$$1 + \frac{1}{n(k_0)} = \left[1 + \frac{1}{n(p_0)}\right]\left[1 + \frac{1}{n(k_0 - p_0)}\right] , \tag{5.106}$$

or equivalently

$$n(k_0) = \frac{n(p_0) n(k_0 - p_0)}{1 + n(p_0) + n(k_0)} . \tag{5.107}$$

It follows that[5]

$$D(p) = \frac{1}{2}\left[\frac{1}{2} + n(p_0)\right] \int \frac{d^4 k}{(2\pi)^4} \left[1 + n(k_0) + n(p_0 - k_0)\right] S(k) S(p - k) . \tag{5.108}$$

On the other hand, the discontinuity becomes

$$\text{Disc } \delta K(p) = -\frac{1}{2} \int \frac{d^4 k}{(2\pi)^4} \left[1 + n(k_0) + n(p_0 - k_0)\right] S(k) S(p - k) , \tag{5.109}$$

after symmetrizing the integral. Therefore,

$$D(p) = -\left[\frac{1}{2} + n(p_0)\right] \text{Disc } \delta K(p) . \tag{5.110}$$

This is a special case of the more general relation

$$\Sigma_{aa}(p) = -\left[\frac{1}{2} + n(p_0)\right] \text{Disc } \Sigma_{ar}(p) , \tag{5.111}$$

where Σ is the self-energy. This formula is ultimately a consequence of the KMS relation discussed earlier. Since we have interpreted Σ_{aa} as the noise correlation

---

[5]Pure vacuum contributions, like $\int S(k) S(p - k)$, are neglected here.

function, while Disc $\Sigma_{ar}(p)$ is the damping rate, we can refer to this relation as the *fluctuation–dissipation theorem*.

If the χ modes are high frequency modes, then it is a good approximation to estimate $D(p) \approx D(0) \equiv D$ and, taking into account the antisymmetry of the imaginary part, Disc $\delta K(p) \approx -2\gamma p_0$. Then we obtain

$$D = 2T\gamma \, , \tag{5.112}$$

which is the Einstein formula for the fluctuation–dissipation theorem.

References

1. Y. Takahashi, H. Umezawa, Thermofield dynamics. Collective Phenom. **2**(55) (1975). (Reprinted: J. Mod. Phys. B **10**, 1755, 1996.)
2. H. Umezawa, *Advanced Field Theory Macro and Thermal Physics* (AIP Press, New York, Micro, 1995)
3. Run-Qin Yang, A complexity of quantum field theory states and application in thermofield double states. Phys. Rev. D **97**, 066004 (2018)
4. R.A. Fisher: *The Negative Binomial Distribution*, Blackwell Publishing Ltd. University College London (1941)
5. M. Joseph, *Hilbe: Negative Binomial Regression* (Cambridge University Press, Cambridge, 2011)
6. H. Abdel El-Shaarawi, Negative binomial Distribution—Applications, in Wiley StatsRef: Statistical Reference Online. https://doi.org/10.1002/9781118445112.stat07353
7. M. Arneodo et al., (EMC): comparison of multiplicity distributions to the negative binomial distribution in muon-proton scattering. Z. Phys. C **35**, 335 (1987)
8. O.G. Tchikilev, Modified negative binomial description of the multiplicity distributions in lepton-nucleon scattering. Phys. Lett. B **388**, 848 (1996)
9. A. Adare et.al. (PHENIX), Charged hadron multiplicity fluctuations in Au + Au and Cu + Cu collisions from $\sqrt{s_{NN}} = 22.5$ to 200 GeV. Phys. Rev. C **78**, 044902 (2008)
10. ALICE Collaboration, K. Aamodt et al., Charged-particle multiplicity measurement in proton–proton collisions at $\sqrt{s} = 0.9$ and 2.36 TeV with ALICE at LHC, EPJ C **68**, 89 (2010)
11. G. Wilk, Z. Wlodarczyk, How to retrieve additional information from the multiplicity distributions. J. Phys. G **44**, 015002 (2017)
12. R. Hagedorn, Nuovo Cimento Suppl. **3**, 147 (1965)
13. R. Hagedorn, Nuovo Cimento A **52**, 64 (1967)
14. R. Hagedorn, Riv. Nuovo Cimento **6**, 1 (1983)
15. J. Rafelski (Ed.), *Melting Hadrons, Boiling Quarks–From Hagedorn Temperature to Ultra-Relativistic Heavy-Ion Collisions at CERN*, SpringerOpen (2016)
16. P. Vilfredo, La courbe de la répartition de la richesse, (Orig. pub., in 1965 *Œuvres complètes de Vilfredo Pareto*, ed. by G. Busino (Librairie Droz, Geneva, 1896)
17. R. Koch, *Living the 80/20 Way: Work Less, Worry Less, Succeed More, Enjoy More* (Nicholas Bearley Pub, London, 2004)
18. W.J. Reed, The Pareto Zipf and other power laws. Econom. Lett. **74**, 15 (2001)
19. C. Tsallis, Nonadditive entropy: the concept and its use. EPJ A **40**, 257 (2009)
20. C.-Y. Wong, G. Wilk, L.J.L. Cirto, C. Tsallis, *From QCD-based hard-scattering to nonextensive statistical mechanical description of transverse momentum spectra in high-energy pp and $p\overline{p}$ collisions*, Phys. Rev. D **91**, 114027 (2015)

21. M. Biyajima, T. Mizoguchi, N. Nakajima, N. Suzuki, G. Wilk, Modified Hagedorn formula including temperature fluctuation—estimation of temperatures at RHIC experiments. EPJ C **48**, 597 (2006)
22. J.M. Zhang, Y. Liu, Fermi's golden rule: its derivation and breakdown by an ideal model. Eur. J. Phys. **37**, 065406 (2016)
23. R.P. Feynman, F.L. Vernon, The theory of a general quantum system interacting with a linear dissipative system. Ann. Phys. **24**, 118 (1963)
24. T.S. Biro, *Is there a temperature? Fundamental Theories of Physics 1014* (Springer, 2011)
25. P.M. Stevenson, Gaussian effective potential: quantum mechanics. Phys. Rev. D **30**, 1712 (1984)
26. P.M. Stevenson, Gaussian effective potential II: $\lambda\varphi^4$ field theory. Phys. Rev. D **32**, 1389 (1985)
27. P.M. Stevenson, Gaussian Effective Potential III: φ^6 theory and bound states. Phys. Rev. D **33**, 2305 (1985)
28. A. Einstein: *Zur Theorie der Brownschen Bewegung*, Ann. Phys. **17**, 549, 1905; **19**, 371, 1906
29. P. Langevin, *Sur la théorie du mouvement brownien*, Comptes Rendus Acad. Sci. (Paris) **146**, 530 (1908)
30. A.D. Fokker, Die mittlere Energie rotierender elektrischer Dipole im Strahlungsfeld. Ann. Phys. **43**, 43 (1914)
31. M. Planck: *Über einen Satz der statistischen Dynamik und seine Erweiterung in der Quantentheorie*, Sitz. Ber. Preuss. Akad. Wiss. **324** (1917)
32. H. Risken, *The Fokker-Planck Equation* (Methods of Solution and Applications, Springer, New York, 1989)
33. C. Greiner: *Interpretation Thermischer Feldtheorie mit Hilfe von Langevin-Prozessen*, (in German), Habilitation thesis, Justus-Liebig University Giessen (1999)
34. E. Cortés, B.J. West, K. Lindenberg, On the generalized Langevin equation: classical and quantum mechanical. J. Chem. Phys. **82**, 2708 (1985)
35. T.S. Biró, G. Purcsel, G. Györgyi, A. Jakovác, Z. Schram, Power-law tailed spectra from equilibrium. Nucl. Phys. A **774**, 845 (2006)
36. T.S. Biro, A. Jakovac, Power-law tails from multiplicative noise. Phys. Rev. Lett. **94**, 132302 (2005)
37. Walter Hans Schottky, Über spontane Stromschwankungen in verschiedenen Elektrizttsleitern. Annalen der Physik **57**, 541 (1918)
38. R. Allard, J. Faubert, D.G. Pelli (eds.), *Using noise to characterize vision* (Frontiers in Psychology, Frontiers Media SA, 2016)

Chapter 6
Maverick Views and Problems

In this last chapter we review a few selected applications of the methods discussed previously. First a realization of an oscillator bath with thermalizing effect is treated for a pure non-Abelian gluon plasma. This is followed by a numerical study of the distribution of local energy packets. Finally, an in-depth study of an exactly solvable model Lagrangian will demonstrate how the spectral function view resolves Gibbs' paradox in a natural way, and how this can be used for a simulation of the equation of state obtained in large scale lattice QCD studies.

6.1 Selected Yang–Mills Plasma Modes

In this section we discuss a physical example which may lead to effects similar to those presented in Sect. 5.3. In fact, non-Abelian plasmas behave analogously in the overdamped situation. The soft equations of motion for such plasmas are strongly reminiscent of classical electrodynamics, the main difference being that the nabla gradient operator is replaced by a gauge-covariant derivative incorporating the non-Abelian vector potential fields. We concentrate on the generation of chromomagnetic fields and work here with units such that $\hbar = 1$ and $c = 1$.

The chromoelectric field in the Hamiltonian gauge is simply $\mathbf{E} = -\dot{\mathbf{A}}$, while the chromomagnetic field is $\mathbf{B} = \mathbf{D} \times \mathbf{A}$. The corresponding Yang–Mills equation reads

$$\mathbf{D} \times \mathbf{B} - \dot{\mathbf{E}} = -\mathbf{J} . \tag{6.1}$$

The source current can take several forms, but in our present example we shall set it to zero. With Ohmic resistance dissipation, it is frequently taken as $\mathbf{J} = -\sigma \mathbf{E}$ in plasmas. In terms of the vector potential fields, the effective equation is then given by

$$\ddot{\mathbf{A}} + \sigma \dot{\mathbf{A}} = \mathbf{D} \times (\mathbf{D} \times \mathbf{A}) . \tag{6.2}$$

In the following, we would like to analyze this equation in terms of Fourier modes of the vector potential, considering su(2) color with three components. In order

© The Author(s), under exclusive licence to Springer Nature Switzerland AG 2019
T. S. Biró and A. Jakovác, *Emergence of Temperature in Examples and Related Nuisances in Field Theory*, SpringerBriefs in Physics,
https://doi.org/10.1007/978-3-030-11689-7_6

to simplify the algebra, we reduce the problem to two nonzero vector potential components: $A_2^1(k) = X(k)$ and $A_1^2(k) = Y(k)$, while the mode vector points in the third direction, viz., $\mathbf{k} = (0, 0, k)$. With this assumption, we obtain the following chromomagnetic field components:

$$\mathbf{B}^1(k) = \left(-kX(k), 0, 0 \right) ,$$
$$\mathbf{B}^2(k) = \left(0, -kY(k), 0 \right) ,$$
$$\mathbf{B}^3(k) = \left(0, 0, -g \sum_{p=0}^{k} X(p)Y(k-p) \right) . \tag{6.3}$$

In the end, we arrive at the following two coupled evolution equations:

$$\ddot{X}(k) + \sigma \dot{X}(k) + k^2 X(k) = g^2 Y(k) \sum_p X(p)Y(k-p) ,$$
$$\ddot{Y}(k) + \sigma \dot{Y}(k) + k^2 Y(k) = g^2 X(k) \sum_p X(p)Y(k-p) . \tag{6.4}$$

Now we wish to distinguish the zero modes from the rest. This yields the following evolution equations for the soft and the hard parts, respectively:

$$\ddot{X}_0 + \sigma \dot{X}_0 = g^2 Y_0 \left[X_0 Y_0 + \sum_{p>0} X(p)Y(-p) \right] ,$$
$$\ddot{Y}_0 + \sigma \dot{Y}_0 = g^2 X_0 \left[X_0 Y_0 + \sum_{p>0} X(p)Y(-p) \right] . \tag{6.5}$$

These feature a multiplicative coupling of the hard modes to the soft fields, besides the non-Abelian anharmonic terms. This is known as the XY-model [1–4]. On the other hand, the hard modes ($k > 0$ only) evolve under the influence of the soft fields:

$$\ddot{X}(k) + \sigma \dot{X}(k) + k^2 X(k) = g^2 Y(k) \left[X_0 Y(k) + X(k)Y_0 + \sum_{0<p<k} X(p)Y(k-p) \right] ,$$
$$\ddot{Y}(k) + \sigma \dot{Y}(k) + k^2 Y(k) = g^2 X(k) \left[X_0 Y(k) + X(k)Y_0 + \sum_{0<p<k} X(p)Y(k-p) \right] . \tag{6.6}$$

This system of equations is reminiscent of those discussed in Sect. 5.4, taking the analytically solvable example of a single soft mode coupled linearly to several oscillators. However, the main effect on the soft pointer degree of freedom, stemming from the hard modes, is not additive, but multiplicative. Multiplicative noise has to be treated with more care than simple additive noise, and it usually leads to a *super-statistical rather than thermal behavior*. Moreover here, reflecting the non-Abelian nature of the underlying forces, the noise couples to the soft degrees of freedom X_0 and Y_0 via an off-diagonal matrix. The noise term, defined as

$$\xi(t) = g^2 \sum_{p>0} X(p,t) Y(-p,t) , \tag{6.7}$$

may have a vanishing expectation value $\langle \xi \rangle = 0$ in most situations, when the two vector potential modes $X(p)$ and $Y(-p)$ are uncorrelated. However, the autocorrelation of the noise already contains the squared amplitudes of the hard modes, and hence is nonzero for nearby times:

$$\langle \xi(t)\xi(t') \rangle = g^4 \sum_{p>0,r>0} \langle X(p,t)X(r,t') \rangle \langle Y(-p,t)Y(-r,t') \rangle \neq 0 . \tag{6.8}$$

As a consequence of the influence of the zero modes on the hard mode dynamics, there is a back-reaction: the autocorrelation properties of the noise do depend on the soft degrees of freedom, and among other things on the energy present in the zero mode. This strengthens the expectation that, in hot non-Abelian plasmas, it is not the simple-minded Boltzmann distribution that dominates. In this model system $\Gamma(E) = \sigma$, and for small k or weak resistance (large σ), this should be a good approximation. On the other hand, $D(E)$ depends on the non-Abelian energy stored in the zero mode.

Finally, we note that not only the zero mode, but an arbitrary set of soft modes could also have been selected for the above study. The qualitative behavior must be the same. But what are the physical properties of the noise and the non-Boltzmannian behavior which are invariant with respect to (or least sensitive to) the precise boundary momentum between the soft and hard modes? Such questions can be addressed in the framework of renormalization studies, a topic elaborated in the next section.

6.2 Distribution of Local Energy Packets

Although big systems can satisfy conditions such that the Gibbs–Boltzmann probability estimate for their energy is an excellent model, local quantities, like energy density in a small volume subsystem, do not necessarily follow the exponentially falling distribution. Even for independent, ideal subsystems in the phase space of simple-minded kinetic models, whenever the subsystem is too small, deviations are expected from the exponential energy distribution. Such deviations are also observed experimentally.

We present here as an example an exploratory study involving numerical and analytical analysis of a classical field theory on a lattice evolving under its own dynamics (closed system dynamics). Here only a microcanonical distribution of energy can be expected, even if equipartition occurs. Once the equipartition of non-interacting kinetic degrees of freedom has been checked, we shall investigate the localized energy density distribution. Since in this study the subsystem is of elementary size, incorporating only a few degrees of freedom, there is no compelling reason to expect an exponential probability density function for this quantity.

In high energy collider experiments, the observed hadron yields are surprisingly far from the exponential energy distribution expected from a naive thermodynamic picture which assumes a canonical state with a sharp value of the temperature. Both at RHIC and LHC, transverse momentum spectra cannot be fit by exponentials, either for heavy-ion or proton–proton collision remnants. Independent measurements in the PHENIX, STAR, CMS, and ALICE collaborations suggest something more like a truncated power-law distribution [5–10].

This distribution has been suggested by Hagedorn [11] as an interpolation formula between the exponential, described as a thermal Boltzmann factor, and a power-law tail, proposed on the basis of perturbative QCD yield calculations. This form,

$$\frac{1}{N} \frac{\mathrm{d}^3 N}{\mathrm{d} p^3} \sim \left(1 + \frac{E}{\nu T}\right)^{-\nu}, \tag{6.9}$$

was recognized as early as the nineteenth century by Vilfredo Pareto in the context of wealth distribution [12–14]. In the late 1980s, a non-Boltzmannian entropy formula was advocated by Constantino Tsallis [15], resulting in the same mathematical form as its canonical energy distribution. This coincidence in the formula opened the possibility to reinterpret and to promote this function as a consequence of a deviation from the additivity of entropy in the systems under study. As pointed out in Chap. 5, the finiteness of the subsystem alone can also result in this form of the energy distribution for the small parts. So the power $\nu = 1/(q-1)$ is a quantified measure of the deviation from the exponential energy distribution which is restored for $q = 1$ ($\nu \to \infty$).

The Tsallis–Pareto fits have only one additional parameter beyond the temperature, describing the shape of the one-particle energy spectra. There are ramifying interpretations of these findings. A QCD-based ansatz can result in parton distribution functions, translated to the hadronic energy distribution in certain hadronization models, which are similar or close to a Tsallis distribution [16, 17]. Essentially simpler reasons can be hidden behind statistical laws of collectivity, such as finite phase space volume effects and the corresponding Gaussian size fluctuations of the temperature. In fact, a gamma distribution of the inverse temperature, $\beta = 1/T$, transforms the exponential energy distribution exactly to a Tsallis distribution [18]. In other models, the fluctuations in the number of kinetic degrees of freedom, and through this the dimensionality of the n-particle phase space, lead to just such a 'superstatistical' distribution of the one-particle energy. In particular, a negative binomial n-particle distribution with a microcanonical phase space factor leads to the Tsallis–Pareto form for this energy [19].

These observations suggest investigating distributions like (6.9) when studying the local energy densities in an extended classical lattice field theory [20]. The observed particle yields depend both on the individual energy distributions of localized sources and on the hadronization mechanism. From a strongly interacting environment, well described in terms of quasiparticle states, messenger particles carry information on localized collective objects (like solitons), and their energy can essentially differ from one-particle energy eigenstates in a free or weakly interacting system. Jet quenching, as observed in relativistic high energy heavy-ion collisions, indicates that the sources of newly created particles are localized in volumes not having a linear size much greater than 1 fm [21–24]. We expect them to carry information rather about the local energy density than about occupation probabilities of global energy levels, which would be the view in thermodynamic approaches. However, these quantities bound to small localities can be influenced by fluctuations very far from the traditional Gaussian assumption.

Local energy densities are numerically measurable in theoretical models with fields. Here we review demonstrations of this for two cases: first a classical Φ^4 field theory and then a quantum SU(3) Yang–Mills theory (with imaginary time). We treat stochastic variables, X, depending on field configurations, A, on the lattice. The indicator function of a local variable is built from Heaviside theta functions:

$$I_X \equiv \Theta(X - x)\Theta(x + \Delta x - X) . \tag{6.10}$$

Its expectation value is the probability that X falls into the interval between x and $x + \Delta x$. On the other hand, canonical thermal averages over field configurations of a quantity $R(A)$ are calculated using functional integrals of the type

$$\langle R(A) \rangle = \int \mathcal{D}A \, R(A) \frac{1}{Z} e^{-\beta H(A)} . \tag{6.11}$$

In numerical simulations, we generate a series of field configurations A_i according to the Boltzmann distribution given above, using the system Hamiltonian $H(A)$. It can either be done by solving the classical field equations (microcanonical approach), or by using Monte Carlo techniques (quantum canonical approach). The expectation values are then estimated as numerical averages, e.g.,

$$\langle R(A) \rangle \approx \frac{1}{N} \sum_{i=1}^{N} R(A_i) . \tag{6.12}$$

In this sense, densities of quantities, in particular the energy density, are obtained from numerical approximations to the expectation value of the corresponding indicator:

$$P(\varepsilon) = \langle \delta(\varepsilon_{X(A)} - \varepsilon) \rangle \tag{6.13}$$

The indicator is a finite range approximation to the Dirac delta distribution in practical numerical approaches.

Before turning to the discussion of numerical simulation results, let us estimate this probability density $P(\varepsilon)$ for free fields. We expand the Dirac delta in the definition of the energy density in Fourier modes:

$$P(\varepsilon) = \int_{-\infty}^{+\infty} \frac{d\lambda}{2\pi} \left\langle e^{i\lambda(\varepsilon_x - \varepsilon)} \right\rangle = \int_{-\infty}^{+\infty} \frac{d\lambda}{2\pi} e^{-i\lambda\varepsilon} \left\langle \sum_{j=0}^{\infty} \frac{(i\lambda\varepsilon_x)^j}{j!} \right\rangle . \tag{6.14}$$

What we need to know here is the expectation value of the jth power of the local energy density, viz., $\langle \varepsilon^j \rangle$, in order to evaluate the above expression. In a free field theory, the energy density can be written in terms of creation and annihilation operators as

$$: \varepsilon_x := \int \frac{d^d p}{(2\pi)^d} \int \frac{d^d q}{(2\pi)^d} \sqrt{\omega_p \omega_q}\, e^{i(p-q)\cdot x}\, a_q^\dagger a_p . \tag{6.15}$$

Here the sign $::$ stands for the normal ordering of creation and annihilation operators. The Hamiltonian is simple in this case:

$$: H := \int \frac{d^d x}{(2\pi)^d} : \varepsilon_x := \int \frac{d^d p}{(2\pi)^d} \omega_p a_p^\dagger a_p . \tag{6.16}$$

Now the expectation value we seek at $x = 0$ is given by

$$\langle : \varepsilon_0^j : \rangle = \frac{1}{Z} \sum_n e^{-\beta E_n} \int \prod_{i=1}^{j} \frac{d^d p_i}{(2\pi)^d} \frac{d^d q_i}{(2\pi)^d} \sqrt{\omega_{p_i} \omega_{q_i}} \langle n| a_{q_1}^\dagger \ldots a_{q_j}^\dagger a_{p_1} \ldots a_{p_j} |n\rangle . \tag{6.17}$$

This expansion is in terms of energy eigenstates, so $E_n = \langle n| : H : |n\rangle$. The operator expectation value is only non-vanishing if the same number of creation and annihilation operators stand in a given term, for each momentum. Cases with more than two operators having equal momenta are suppressed by factors of the total volume in the d-dimensional phase space, so we can neglect them. This is the free field version of the thermodynamic limit. Then, due to Wick's theorem, all simple pairings give the same contribution and we obtain

$$\langle n| a_{q_1}^\dagger \ldots a_{q_j}^\dagger a_{p_1} \ldots a_{p_j} |n\rangle = j! \prod_{i=1}^{j} n(p_i)\, (2\pi)^d \delta(p_i - q_i) . \tag{6.18}$$

This implies

$$\langle : \varepsilon_0^j : \rangle = j! \langle : \varepsilon_0 : \rangle^j . \tag{6.19}$$

From this, we arrive at the result

$$\langle\, e^{i\lambda\varepsilon_x}\,\rangle = \sum_{j=0}^{\infty}(i\lambda)^j\langle\,\varepsilon_x\,\rangle^j = \frac{1}{1 - i\lambda\langle\,\varepsilon_x\,\rangle}\,. \tag{6.20}$$

This expression has a pole at $\lambda_{\text{pole}} = -i/\langle\,\varepsilon_x\,\rangle$, so the probability distribution function of the energy density can be calculated from the λ-integral (6.14) using the residue at this pole[1]:

$$P(\varepsilon) = \frac{1}{\langle\,\varepsilon\,\rangle}\,e^{-\varepsilon/\langle\,\varepsilon\,\rangle}\,. \tag{6.21}$$

This is an exponentially decreasing probability of having an energy density above a given value, and reminds us of the Boltzmann–Gibbs thermodynamic factor. Equipartition is realized via the temperature parameter (inverse slope), which is the expectation value of the localized energy packets.

The occurrence of an exponentially falling probability for the local energy density is guaranteed for a free field theory in the thermodynamic limit, as the calculations above clearly demonstrate. Reversing this argument, whenever a non-exponential distribution of this quantity is discovered, we may suspect that non-trivial correlations between local energy densities play a role in the background. This may stem from having too small a total volume (an assumption extraneous to traditional field theory) or from a non-trivial and strong coupling between the elementary fields.

We now consider the classical Φ^4 theory. The Hamiltonian is a sum of local energy density expressions:

$$\varepsilon_x = \frac{1}{2}\Pi_x^2 + \frac{1}{2}(\nabla\Phi_x)^2 + \frac{m^2}{2}\Phi_x^2 + \frac{g^2}{4!}\Phi_x^4\,. \tag{6.22}$$

On a lattice with spacing a, we use discretized gradients $\nabla_i\Phi_x = \Phi_{x+ae_i} - \Phi_x$. The Hamiltonian equations of motion are

$$\dot{\Phi}_x = \Pi_x\,, \qquad \dot{\Pi}_x = \nabla^2\Phi_x - m^2\Phi_x - g^2\Phi_x^3/3!\,. \tag{6.23}$$

Energy is conserved by these equations, but solving in discretized time, higher order deviations may occur. The equations of motion can be rescaled by $t \to t/a$, together with $\Phi \to ga\Phi$, $\Pi \to ga^2\Pi$, $\nabla = a\nabla$, and $m \to am$. In the scaled quantities, the classical equations of motion become independent of the scalar self-coupling:

$$\dot{\Phi}_x = \Pi_x\,, \qquad \dot{\Pi}_x = \nabla^2\Phi - m^2\Phi - \frac{1}{6}\Phi^3\,. \tag{6.24}$$

The energy density scales according to $\varepsilon_x \to g^2a^4\varepsilon_x$.

In numerical simulations with a time step of $0.01a$, after about twenty steps, a Tsallis–Pareto distribution is established which holds for a long time, in fact, over

---

[1]Normalized so that $\int_0^{\infty} P(\varepsilon)d\varepsilon = 1$.

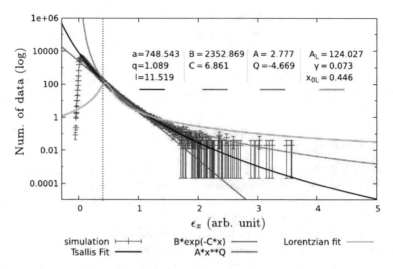

Fig. 6.1 Distribution of energy density in classical Φ^4 lattice field theory after a fast initial evolution. Each point is an average of 50 simulations. From [20]

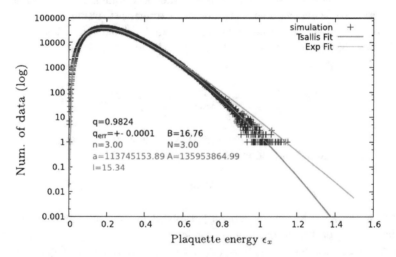

Fig. 6.2 Plaquette energy distribution from an SU(3) Yang–Mills lattice simulation in four dimensions

five hundred time steps. Of course, numerically, several other functional forms may fit, but this was found to be the best approximation (see Fig. 6.1).

Another numerical simulation for a quantum SU(3) Yang–Mills theory on a lattice led to the one-plaquette energy distribution. Figure 6.2 presents the histogram distribution.

For renormalization of the Tsallis parameter q, obtained from fitting the truncated power-law energy density distribution, the Sommer scale parameter $r_0 \approx 0.5$ fm has

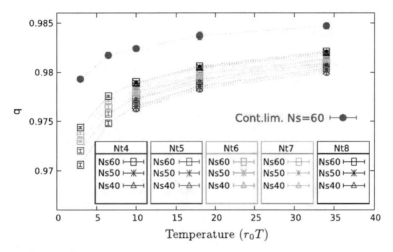

Fig. 6.3 Tsallis parameter q as a function of the scaled temperature, $r_0 T$, for different size lattices, and its continuum limit extrapolation

been used for its dependence on the dimensionless temperature $r_0 T$. Results are plotted in Fig. 6.3.

Summarizing this section, numerical determination of classical Φ^4 and quantum SU(3) Yang–Mills field theories reveals local energy density probability distributions which are not exponential, in spite the thermal conditions set by a continuum limit analysis for the global systems (defining the path of constant physics). The results are best fitted by Pareto–Hagedorn–Tsallis distributions, but with different values of the parameter q. For the Φ^4 simulation, values around $q = 1.024$, and for the Yang–Mills simulation, values around $q = 0.983$ are obtained as continuum field theory extrapolations. It would also be interesting to look for nontrivial correlations caused by field theoretical interactions in other, possibly analytically tractable models, in order to find out what determines the actual value of q and how it approaches 1, if at all.

6.3 Melting of Spectra

At the beginning of quark–gluon plasma (QGP) research [25], based on the MIT bag model [26], it was assumed that the transition between hadronic matter and QGP was like ice melting into water. A latent heat was assumed, and a sharp transition line in the plane of the temperature (T) and the baryochemical potential (μ). More realistically, numerical solutions to the full QCD problem on Euclidean lattices have since shown that, at zero and low μ values, this transition is a crossover [27], more reminiscent of the way butter melts in the sun, and it happens without any latent heat. This is not a trivial question, since with static (infinitely heavy) quarks, the transition would be with latent heat, related to a jump in the energy density at a sharp temperature value [28, 29].

In this section we introduce an analytically solvable field theoretical model (hence of academic interest), and study the melting of spectra and some other essential thermodynamic properties of media with known spectral functions. We use units such that $\hbar = c = k_B = 1$. We study a theory quadratic in the fields $\varphi(x)$, but with a general kernel in the action [30]:

$$S[\varphi] = \frac{1}{2} \int d^4x \varphi(x) \mathcal{K} (i\partial) \varphi(x) . \tag{6.25}$$

Expanded in a plane wave basis, we have the equivalent definition

$$S[\varphi] = \frac{1}{2} \int \frac{d^4 p}{(2\pi)^4} \varphi^*(p) \mathcal{K} (p) \varphi(p) . \tag{6.26}$$

Here we use the same notation for the φ field in both the x- and the p-representation.

The scalar field $\varphi(x)$ is real, so $\varphi^*(p) = \varphi(-p)$. To obtain a real action, the kernel must obey the relations $\mathcal{K}^*(p) = \mathcal{K}(-p) = \mathcal{K}(p)$. The retarded propagator is the inverse of the kernel in this representation, with the corresponding pole prescription on the complex energy plane:

$$G^{ra}(p) = \mathcal{K}^{-1}(p_0 + i\varepsilon, \mathbf{p}) , \tag{6.27}$$

where ε is an arbitrarily small, but positive quantity. According to the general rule, the spectral function is the discontinuity when crossing the real axis in the energy plane. This is an application of (2.75) with $A = B = \varphi$:

$$S(p) = i G^{ra}(p) - i G^{ar}(p) . \tag{6.28}$$

Its inverse relation follows from the Kramers formula (2.74):

$$\mathcal{K}(p) = \Re \left(\int \frac{d\omega}{2\pi} \frac{S(\omega, \mathbf{p})}{p_0 - \omega + i\varepsilon} \right)^{-1} . \tag{6.29}$$

Now, a field theory defined by an arbitrary spectral function $S(p)$ is not necessarily consistent. However, starting from a spectral function with the property $S(-p_0, \mathbf{p}) = -S(p_0, \mathbf{p})$, using the $\Re \, p_0 > 0$ half-plane for the definition, the kernel $\mathcal{K}(p)$ obtained from the above equation defines an action for a field theory consistent with the real field $\varphi(x)$. If in addition the spectral function is Lorentz invariant, so will be the model field theory. Given such a theory, we are seeking the thermodynamic behavior it describes.

Standard techniques like the Hamiltonian approach do not work in this case: the time nonlocality inherent in the p_0-dependence of $\mathcal{K}$ prevents definition of a conjugate momentum $\Pi(x)$. Instead, we build the thermodynamic treatment on the energy density and pressure contained in the energy–momentum tensor $T_{\mu\nu}(x)$ derived from the above action (6.25). The energy density is the time–time component, T_{00}, while the total tensor represents the response to changing the coordinates by a spacetime translation

$$x'_\mu = x_\mu + a_\mu(x) \ . \tag{6.30}$$

The scalar field does not change under this transformation, $\varphi'(x') = \varphi(x)$. The change in the action due to this transformation is

$$S[\varphi'] - S[\varphi] = -\int d^4x \, T_{\mu\nu}(x)\partial^\mu a^\nu \ , \tag{6.31}$$

to leading (linear) order in the small but space- and time-dependent shift. Under the local shift (6.30), the coordinate differentials transform as

$$dx'_\mu = dx_\mu + (\partial_\mu a^\nu)dx_\nu \ , \tag{6.32}$$

and the partial derivatives contravariantly, i.e.,

$$\partial'_\mu = \partial_\mu - (\partial_\mu a^\nu)\partial_\nu \ , \tag{6.33}$$

to leading order. This leads to extra terms when we expand $\mathscr{K}(i\partial)$.

The action in the shifted coordinates reads

$$S[\varphi'] = \frac{1}{2}\int d^4x' \, \varphi'(x')\mathscr{K}(i\partial')\varphi'(x') \ , \tag{6.34}$$

which can now be expressed in terms of the original field, coordinates, and partial derivatives:

$$S[\varphi'] = \frac{1}{2}\int d^4x \, \left(1 + \partial_\lambda a^\lambda\right)\varphi(x)\mathscr{K}\left(i\partial_\mu - (\partial_\mu a^\nu)i\partial_\nu\right)\varphi(x) \ . \tag{6.35}$$

Collecting the contributions to leading order in the shift $a(x)$, we arrive at the following energy–momentum tensor:

$$T_{\mu\nu}(x) = \frac{1}{2}\varphi(x)\mathscr{D}_{\mu\nu}(i\partial)\varphi(x) \ , \tag{6.36}$$

with

$$\mathscr{D}_{\mu\nu}(i\partial) = \left(\nabla_\mu\mathscr{K}\right)\cdot i\partial_\nu - g_{\mu\nu}\mathscr{K} \ . \tag{6.37}$$

Here the ∇ operator notation is the formal derivative of the defining kernel $\mathscr{K}$ with respect to its corresponding argument. The Fourier transform of the energy–momentum kernel is simpler. In an index-symmetrized version (the one in which it will be used), it is given by

$$\mathcal{D}_{\mu\nu}(p) = \frac{1}{2}\left(p_\mu \frac{\partial \mathcal{K}}{\partial p^\nu} + p_\nu \frac{\partial \mathcal{K}}{\partial p^\mu}\right) - g_{\mu\nu}\,\mathcal{K}(p)\;. \tag{6.38}$$

Once this kernel is obtained, we can proceed to calculate the expectation values and equation of state, or transport coefficients. The Fourier space energy–momentum tensor thus becomes

$$T_{\mu\nu}(k) = \frac{1}{2}\int \frac{\mathrm{d}^4 p}{(2\pi)^4}\,\varphi(k-p)\mathcal{D}_{\mu\nu}(p)\varphi(p)\;. \tag{6.39}$$

Its expectation value in a condensate or in a homogeneous medium satisfying

$$\langle\,\varphi(k-p)\varphi(p)\,\rangle = \delta(k)\langle\,\varphi(-p)\varphi(p)\,\rangle \tag{6.40}$$

is proportional to $\delta(k)$, and in coordinate space, it is independent of the position:

$$T_{\mu\nu}^{\mathrm{hom}}(k) = \frac{1}{2}\delta(k)\int \frac{\mathrm{d}^4 p}{(2\pi)^4}\,\mathcal{D}_{\mu\nu}(p)\langle\,\varphi(-p)\varphi(p)\,\rangle. \tag{6.41}$$

The boson field propagator described by the above expectation value is composed of the Bose distribution and the spectral function (as discussed in previous sections). Considering also the symmetry between positive and negative energy states (KMS relation), we arrive at the result

$$T_{\mu\nu}^{\mathrm{hom}}(k) = \delta(k)\int_0^\infty \frac{\mathrm{d}p_0}{2\pi}\int \frac{\mathrm{d}^3\mathbf{p}}{(2\pi)^3}\left[n(p_0) + \frac{1}{2}\right]S(p)\mathcal{D}_{\mu\nu}(p)\;. \tag{6.42}$$

The 00 component of this result delivers the energy density, and its four-trace the quantity $\varepsilon - 3p$ in an isotropic medium.

The expression (6.42) is a nonlinear functional of the spectral density $S(p)$. It is constructed in such a way that a constant factor cancels; a wave function renormalization causing $G \to ZG$ and thereby attaching the same factor to the spectral density, viz., $S \to ZS$, cancels due to $\mathcal{D}_{\mu\nu} \to \mathcal{D}_{\mu\nu}/Z$, which follows from $\mathcal{K} \to \mathcal{K}/Z$ [see (6.27)]. The factor $1/2$ occurring with the energy-dependent occupation number in equilibrium, viz., $n(p_0)$, is a vacuum contribution. This can be disregarded by assuming that a proper renormalization has already been carried out at zero temperature.

The renormalized energy density $\varepsilon = u^\mu u^\nu T_{\mu\nu}$ contains the kernel

$$u^\mu u^\nu \mathcal{D}_{\mu\nu}(p) = p_0 \frac{\partial \mathcal{K}}{\partial p_0} - \mathcal{K}\;, \tag{6.43}$$

while the isotropic hydrostatic pressure, obtainable from $\mathcal{D}_\lambda{}^\lambda(p) = p^\mu \nabla_\mu \mathcal{K} - 4\mathcal{K}$, contains the factor

$$\mathcal{P}(p) = \mathcal{K} - \frac{1}{3}\mathbf{p}\cdot\frac{\partial}{\partial\mathbf{p}}\mathcal{K}\;. \tag{6.44}$$

We thus arrive at the following expressions for the energy density, pressure, and entropy density, respectively:

$$\varepsilon = \iint \mathscr{D}_{00} \, ,$$

$$P = \iint \frac{1}{3} \left(\mathscr{D}_{00} - \mathscr{D}_{\mu}{}^{\mu} \right) \, ,$$

$$s = \frac{1}{T} \iint \left(\frac{4}{3} \mathscr{D}_{00} - \frac{1}{3} \mathscr{D}_{\mu}{}^{\mu} \right) \, , \tag{6.45}$$

with the following short-hand notation for a four-fold momentum integral weighted by the spectral function and the equilibrium energy level occupation Bose factor:

$$\iint \ldots = \int \frac{d^3\mathbf{p}}{(2\pi)^3} \int_0^\infty \frac{d p_0}{2\pi} S(p_0, \mathbf{p}) n(p_0/T) \ldots \, . \tag{6.46}$$

Let us review a few simple cases. A traditional free scalar field theory is defined by $\mathscr{K} = p_\mu p^\mu$. In this case, we obtain

$$\mathscr{D}_{\mu\nu} = 2 p_\mu p_\nu - g_{\mu\nu} (p_\lambda p^\lambda) \, ,$$

$$\mathscr{D}_{00} = p_0^2 + |\mathbf{p}|^2 \, ,$$

$$\mathscr{D}_\mu{}^\mu = 2|\mathbf{p}|^2 - 2 p_0^2 \, . \tag{6.47}$$

The resulting thermodynamic quantities are then

$$\varepsilon = \iint \left(p_0^2 + |\mathbf{p}|^2 \right) \, ,$$

$$P = \iint \left(p_0^2 - \frac{1}{3} |\mathbf{p}|^2 \right) \, ,$$

$$s = \frac{1}{T} \iint \left(2 p_0^2 + \frac{2}{3} |\mathbf{p}|^2 \right) \, . \tag{6.48}$$

For such a $\mathscr{K}$, the spectral density – calculated from the imaginary part of the retarded propagator, the inverse of $\mathscr{K}$ – becomes

$$S(p) = 2\pi \delta \left(p_\lambda p^\lambda \right) \, , \tag{6.49}$$

and all integrals over p_0 can be carried out. Naturally, this delivers the known results for massless bosons:

$$\varepsilon = \int \frac{d^3\mathbf{p}}{(2\pi)^3} |\mathbf{p}| n(|\mathbf{p}|/T) \, ,$$

$$P = \int \frac{d^3\mathbf{p}}{(2\pi)^3} \frac{1}{3} |\mathbf{p}| n(|\mathbf{p}|/T) \, ,$$

$$s = \frac{1}{T} \int \frac{d^3\mathbf{p}}{(2\pi)^3} \frac{4}{3} |\mathbf{p}| n(|\mathbf{p}|/T) \, . \tag{6.50}$$

A further test of these concepts is given by investigating spectral functions consisting of Dirac delta peaks at well-defined, separate energies. In [31–33] it is argued that a general sum of such terms also leads to additive thermodynamics. Starting with a propagator as a sum of weighted poles,

$$G = \sum_i \frac{Z_i}{p_0 - \omega_i(\mathbf{p}) + i\varepsilon} \, , \tag{6.51}$$

the corresponding spectral function is a sum of sharp spectral lines,

$$S = \sum_i Z_i 2\pi \delta(p_0 - \omega_i) \, . \tag{6.52}$$

In this case, when calculating thermodynamic quantities, only the pole contributions count. Since $\mathscr{K} = G^{-1}$, we have

$$p_0 \frac{\partial \mathscr{K}}{\partial p_0} = -\frac{p_0}{G^2} \frac{\partial G}{\partial p_0} = p_0 \frac{\displaystyle\sum_j \frac{Z_j}{\left(p_0 - \omega_j\right)^2}}{\left(\displaystyle\sum_i \frac{Z_i}{p_0 - \omega_i}\right)^2} \, . \tag{6.53}$$

In the neighborhood of a given pole, say $p_0 \to \omega_i$, the sums are dominated by that single pole contribution. Taking such a limit, we conclude that

$$\lim_{p_0 \to \omega_i} p_0 \frac{\partial \mathscr{K}}{\partial p_0} = \frac{\omega_i}{Z_i} \, , \tag{6.54}$$

and as a consequence

$$\mathscr{D}_{00} \longrightarrow \sum_i 2\pi Z_i \delta(p_0 - \omega_i) \frac{\omega_i}{Z_i} = \sum_i \omega_i 2\pi \delta(p_0 - \omega_i) \tag{6.55}$$

contributes to the energy density from each pole simply with the corresponding energy ω_i.

The question of the effective number of degrees of freedom also arises in this description. In the above extreme situation, it should be the number of poles, but what about a more general definition? Physically, we are looking for a concept based on the energy density expression, akin to the Williams–Weiszäcker view of the number of quanta in a field. We would also like to base our definition on a temperature-independent founding. Moreover, the factors in the propagator terms, sensitive to the wave function renormalization, have to be irrelevant.

Such a proposition was made in [31]. The proposed definition, viz.,

$$N_{00} \equiv \int_0^\infty \frac{dp_0}{2\pi} \frac{1}{p_0} \mathscr{D}_{00}(p_0, \mathbf{p}) S(p_0, \mathbf{p}) , \qquad (6.56)$$

involves the spectral function and the proper constructive term from the Lagrangian, but omits the thermal equilibrium distribution. For the sum of spectral lines discussed above, this is simply the number of lines:

$$N_{00}^{poles} = \int_0^\infty \frac{dp_0}{2\pi} \sum_{i=1}^N \frac{2\pi \omega_i}{p_0} \delta(p_0 - \omega_i) = N . \qquad (6.57)$$

This simplicity changes if we consider finite width resonances. The relativistic propagator for a single finite width resonance is

$$G = \frac{1}{(p_0 + i\Gamma)^2 - \omega^2} = \frac{1}{2\omega} \left(\frac{1}{p_0 + i\Gamma - \omega} - \frac{1}{p_0 + i\Gamma + \omega} \right) . \qquad (6.58)$$

The corresponding spectral function $S = i(G - G^*)$ becomes

$$S = \frac{4\Gamma p_0}{\left[(p_0 - \omega)^2 + \Gamma^2\right]\left[(p_0 + \omega)^2 + \Gamma^2\right]} . \qquad (6.59)$$

The identification $\mathscr{K} = \Re G^{-1} = p_0^2 - \omega^2 - \Gamma^2$ leads to

$$\mathscr{D}_{00} = p_0^2 + \omega^2 + \Gamma^2 . \qquad (6.60)$$

The effective number of degrees of freedom, based on the definition (6.56), is given by the integral

$$N_{00}^\Gamma = \int_0^\infty \frac{dp_0}{2\pi} \frac{4\Gamma \left(p_0^2 + \omega^2 + \Gamma^2\right)}{\left[(p_0 - \omega)^2 + \Gamma^2\right]\left[(p_0 + \omega)^2 + \Gamma^2\right]} . \qquad (6.61)$$

The integrand is easily recognized as a sum of two reciprocals:

$$N_{00}^\Gamma = \int_0^\infty \frac{dp_0}{2\pi} 2\Gamma \left[\frac{1}{(p_0 - \omega)^2 + \Gamma^2} + \frac{1}{(p_0 + \omega)^2 + \Gamma^2} \right] . \qquad (6.62)$$

This expression for the integrand is an even function of p_0, so the integral can be extended to the interval $[-\infty, +\infty]$. After this extension, the shift of $\pm\omega$ does not matter any more, and we arrive at a sum of two elementary integrals:

$$2N_{00}^{\Gamma} = \int\limits_{-\infty}^{+\infty} \frac{dp_0}{2\pi} 2\Gamma \left[\frac{1}{(p_0 - \omega)^2 + \Gamma^2} + \frac{1}{(p_0 + \omega)^2 + \Gamma^2} \right] , \qquad (6.63)$$

resulting in

$$2N_{00}^{\Gamma} = \frac{\Gamma}{\pi} \left(\frac{\pi}{\Gamma} + \frac{\pi}{\Gamma} \right) = 2 . \qquad (6.64)$$

This implies $N_{00}^{\Gamma} = 1$, independently of the width Γ, as expected.

The entropy, energy density, and pressure can only be obtained numerically, since the integrals involve the Bose distribution $n(p_0/T)$. Results of such calculations are shown in Fig. 6.4, considering a unit mass and different resonance widths.

The next interesting case is the study of two finite width resonances. When the masses of the resonances are too close to each other, any distinction between them as individual particles diminishes. The most extreme manifestation of the underlying problem is given in the Gibbs paradox [33, 34]: as long as two particles have different masses, their ideal gas mixture will have twice as many degrees of freedom than if they had equal mass. The transition looks like a sudden halving of the number of degrees of freedom at mass equality.

Finite width resonances must resolve this sudden behavior. We expect *the melting of the two resonances to one broad spectrum to be continuous*. However, it is not enough to study the mixed spectral function, deciding whether it has one or two maxima; for the thermodynamic behavior – as discussed above – the study of $\mathcal{D}_{00}(p_0, \mathbf{p})$ is also of relevance.

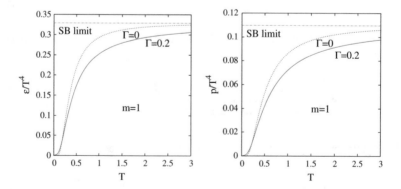

Fig. 6.4 The T^4-normalized energy density (*left*) and pressure (*right*) of a single resonance spectral function, at unit mass $\omega(0) = m = 1$, with zero and nonzero widths. SB indicates the Stefan–Boltzmann limit

A linear mixture of two retarded propagators can be written

$$G = Z_1 G_1 + Z_2 G_2 = \frac{Z_1}{a_1 b_1} + \frac{Z_2}{a_2 b_2} \ . \tag{6.65}$$

Here we introduce the notation $a = p_0 - \omega - i\Gamma$ and $b = p_0 + \omega - i\Gamma$. Then $b - a = 2\omega$ and the individual spectral functions $S = 2\,\Im m\, G$ mix as $S = Z_1 S_1 + Z_2 S_2$. In the derivation of the corresponding $\mathscr{D}_{00}$, on the other hand, terms occur which contain nonlinear information about the mixing. We start by noting the derivative of the inverse propagator:

$$\frac{\partial}{\partial p_0} G^{-1} = -\frac{1}{(Z_1 G_1 + Z_2 G_2)^2} \left(Z_1 \frac{\partial G_1}{\partial p_0} + Z_2 \frac{\partial G_2}{\partial p_0} \right) \ . \tag{6.66}$$

On the other hand, the individual derivatives of propagators are given by

$$\frac{\partial G_i}{\partial p_0} = \frac{\partial}{\partial p_0} \frac{1}{a_i b_i} = -2(p_0 - i\Gamma_i) G_i^2 \ . \tag{6.67}$$

For such a mixture of two resonances, we obtain the following complex expression

$$\mathscr{D}_{00} = p_0 \frac{\partial}{\partial p_0} G^{-1} - G^{-1} = \frac{\left[2p_0(p_0 - i\Gamma_i) Z_1 G_1^2 - Z_1 G_1 \right] + [1 \leftrightarrow 2]}{(Z_1 G_1 + Z_2 G_2)^2} \ . \tag{6.68}$$

Here the $1 \leftrightarrow 2$ notation in the square bracket in the numerator indicates exchange of the indices 1 and 2. Now factorizing out the $Z_i G_i^2$ in the square brackets, the remaining part becomes

$$R \equiv 2p_0(p_0 - i\Gamma) - ab = p_0^2 + \Gamma^2 + \omega^2 \ . \tag{6.69}$$

This leads to the final form

$$\mathscr{D}_{00} = \frac{Z_1 R_1 G_1^2 + Z_2 R_2 G_2^2}{(Z_1 G_1 + Z_2 G_2)^2} \ . \tag{6.70}$$

This expression is still complex and can be written in the alternative form

$$\mathscr{D}_{00} = \frac{Z_1 R_1 + Z_2 R_2 \xi^2}{(Z_1 + Z_2 \xi)^2} \ , \tag{6.71}$$

with

$$\xi \equiv \frac{G_2}{G_1} = \frac{a_1 b_1}{a_2 b_2} \ . \tag{6.72}$$

Note that $\xi(-p_0) = \xi^*(p_0)$, so $\mathscr{D}_{00}$ has the same property. This is important when calculating the number of degrees of freedom. The integration over p_0 can be extended to the interval $[-\infty, +\infty]$ using the real part of $\mathscr{D}_{00}$. Furthermore, it is interesting to note that (6.70) is equivalent to an expression using the inverse propagators:

$$\mathscr{D}_{00} = \frac{Z_1 R_1 (a_2 b_2)^2 + Z_2 R_2 (a_1 b_1)^2}{(Z_1 a_2 b_2 + Z_2 a_1 b_1)^2} . \tag{6.73}$$

Here the denominator can be interpreted as the square of an effective polynomial:

$$Z_1 a_2 b_2 + Z_2 a_1 b_1 = p_0^2 - \omega_{12}^2 - 2i p_0 \Gamma_{12} , \tag{6.74}$$

with

$$\omega_{12}^2 \equiv Z_1 (\omega_2^2 + \Gamma_2^2) + Z_2 (\omega_1^2 + \Gamma_1^2) \tag{6.75}$$

and

$$\Gamma_{12} \equiv Z_1 \Gamma_2 + Z_2 \Gamma_1 . \tag{6.76}$$

Unfortunately, the numerator in (6.73) cannot be reinterpreted in such an easy way. It remains to study the mixtures numerically with respect to degrees of freedom and energy density, pressure, and other thermodynamic quantities.

For the energy density and pressure, we present the results of numerical calculations in Fig. 6.4. As a reference, we show the zero width case (two Dirac delta peaks in the spectral function). Note that, even for a width of $\Gamma = 0.2\,m$ [at zero momentum $\omega(0) = m$], the energy density and pressure curves are close to the free case, in particular for low energies. For this reason, the hadron resonance gas model [35–39] of the interacting quark–gluon plasma is successful in this regime of width values, characteristic, for instance, of the rho meson.

The spectral function as a sum of two Lorentzians, $S = 2\,\mathfrak{Im}\,G = Z_1 S_1 + Z_2 S_2$, is shown in Fig. 6.5. Considering the number of degrees of freedom, the traditional view would count 2 for two finite width particles in cases when the spectral function is a sum of two Lorentzians ($Z_1 = Z_2 = 1$). The energy density and pressure are

Fig. 6.5 Spectral function as a sum of two Lorentzians with the same width ($\Gamma = 0.2$) and masses $m = 1$ and $m = 2$. *Dashed lines* show the individual parts of the spectral function

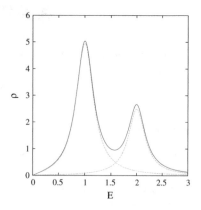

Fig. 6.6 Number of degrees of freedom at fixed unequal masses, but with varying widths of the Lorentzian peaks. The *solid line* is obtained using the definition and the *dashed line* using an approximate fit

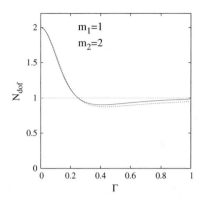

supposed to be simply the sum of the individual contributions. In contrast, in the one-field description presented in this section, we obtain a more complex result. Using the expressions for $\mathcal{D}_{00}(p_0, \mathbf{p})$ and $S(p_0, \mathbf{p})/p_0$ presented above, the energy density and pressure can be obtained numerically, while $N_{00}^{\Gamma_1, \Gamma_2}$ can be derived using the analytical preparatives discussed above.

For the choice $m_1 = 1$, $m_2 = 2$, the number of degrees of freedom is plotted in Fig. 6.6. In order to gain a better insight into these formulæ, it is worth working out an approximation to the function N_{dof} with the variable Γ, for $|m_1 - m_2| \ll m_{1,2}$. We obtain

$$N_{\mathrm{dof}} = 1 + \frac{1 - y^2}{(1 + y^2)^2}, \quad \text{where} \quad y = \frac{4\Gamma}{|m_2 - m_1|}. \tag{6.77}$$

Although this formula is just a wild guess on the basis of a few trials, it provides a good approximation to the correct expression.

It is a beautiful feature of our definition of the number of degrees of freedom that, for narrow widths, it starts at 2: this is correct for two independent Dirac deltas in the spectrum. This is a special case of (6.57) when $N = 2$. On the other hand, as $\Gamma \to \infty$, this number approaches one. This can also be investigated by means of analytic calculations. For $\Gamma_1 \to \infty$ and $\Gamma_2 \to \infty$, we can use a common value for the widths, i.e., $\Gamma_1 \to \Gamma$ and $\Gamma_2 \to \Gamma$, and let $\Gamma \to \infty$ at the end. In this case, we may also take the limit $\omega_i \to 0$, while p_0 has to be left unlimited because of the integration over this variable. In this limit, $\xi \to 1$ and

$$\mathcal{D}_{00} \longrightarrow \frac{p_0^2 + \Gamma^2}{Z_1 + Z_2}. \tag{6.78}$$

In this limit, the other factor in the integrand of the integral determining N_{00} is

$$S/p_0 \longrightarrow \frac{4\Gamma}{(p_0^2 + \Gamma^2)^2}(Z_1 + Z_2). \tag{6.79}$$

This gives rise to the result

$$N_{00}^{\Gamma_i \to \infty} = \int\limits_0^\infty \frac{\mathrm{d}p_0}{2\pi} \frac{4\Gamma}{p_0^2 + \Gamma^2} = 1 \,. \tag{6.80}$$

The two peaks merge, and there is no way of guessing that there are two separate peaks. This phenomenon and its counterpart with several peaks is called *melting of resonances*. In practical physical situations, in going from the $\omega \gg \Gamma$ to the $\omega \ll \Gamma$ case, the effective number of degrees of freedom N_{00} reduces from N to 1. Using this technique we have changed the system with two degrees of freedom to one with one degree of freedom without deliberately changing the number of field components.

In situations with realistic widths and overlaps between resonance peaks, the number of degrees of freedom is not physically defined and the integral for N_{00} can have fractional values. *There can even be regions with $N_{00} < 1$*, which may indicate an 'over-binding' between the individual peaks in the original spectral function. In our example depicted in Fig. 6.5, the two peaks are still individually identifiable, while the function N_{00} given by our definition is already close to 1.

This phenomenon is closely related to the resolution of the Gibbs paradox. This arises when, in a system with two degrees of freedom, we continuously decrease the 'difference' between the peak parameters; then for any nonzero difference, a system with two degrees of freedom is observed, which abruptly changes to one with one degree of freedom at zero difference. The resolution of this paradox is that this change is in fact smooth, if we consider finite lifetime particles. It seems that physically we cannot perform a sufficiently long measurement which is accurate enough to distinguish the two separate spectral peaks when the resonance lifetimes get too short.

Although our construction of the integral quantity N_{00} provides a good description of the essential physics of this change, we may still be interested in the fate of observable thermodynamic quantities. As a guide, we numerically determined the energy density and pressure integrals with Bose distributions for two resonances. The results are plotted in Fig. 6.7.

The curve labeled (iii) represents the exact formulas for ε/T^4 and p/T^4 when $\Gamma = 0.2$. In an additive picture of thermodynamic contributions from the individual Lorentzians, one would obtain curve (ii). As a baseline reference, the zero width case is shown by curve (i). Finally, curve (iv) presents the result for a single Lorentzian with the parameters $m = 1.2$ and $\Gamma = 0.2$. It is clearly demonstrated that *the independent resonance approximation fails for two overlapping Lorentzians.* While the former is close to the result with two Dirac delta peaks, the exact result is more like the common single Lorentzian. The thermodynamic study therefore supports the conclusion that *the two peak system with a sizeable width behaves like an object with one degree of freedom.*

Besides considering a resonance gas, there are further characteristic spectral functions beyond the Dirac delta peaks. We now choose a single threshold spectral function

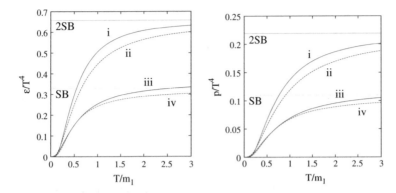

Fig. 6.7 $1/T^4$ normalized energy density (*left*) and pressure (*right*) for a spectral function constructed as a sum of two Lorentzian peaks at $m_1 = 1$, and $m_2 = 2$, respectively, with the same width Γ. The curves for $\Gamma = 0.2$ are labelled (iii) in both figures. Some other curves are shown for comparison: (i) for the two Dirac delta case, (ii) for the independent resonance approximation, (iv) for a single Lorentzian around the mass $m = 1.2$, with width $\Gamma = 0.2$. SB indicates the Stefan–Boltzmann limit, and 2SB twice the Stefan–Boltzmann limit

Fig. 6.8 Spectral function of a single threshold

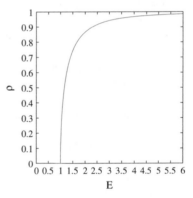

$$\rho(p) = \theta(p^2 - M^2)\sqrt{1 - \frac{M^2}{p^2}}, \qquad (6.81)$$

as depicted in Fig. 6.8. In principle this spectral function can be represented with some accuracy by a number of massively overlapping Lorentzian peaks. Since normalization factors drop out, even the relative weights can be chosen with some freedom. The threshold behavior, however, is far from any particle-like spectral function: no peak occurs and it does not diminish at infinite energy. Such spectral functions usually arise as parts of strongly correlated systems, where particle creation and annihilation is an abundant physical process. Here we handle it as a mathematical example to explore the phenomenon of melting in another extreme form, as a merger with the continuum.

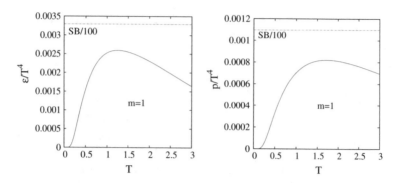

Fig. 6.9 Energy density and pressure for the single threshold spectral function. SB/100 stands for one hundredth of the Stefan–Boltzmann limit

When we evaluate the number of degrees of freedom using the definition (6.56), we obtain $N_{00} = 0.98$ for $m = 1$ in the threshold formula (6.81). Curves for the pressure and energy density were obtained numerically again, as shown in Fig. 6.9. The most striking feature of this result is its amplitude: the contribution to the energy density and pressure of branch cuts causing threshold behavior in the spectral function is very small, although the contribution to the integral giving the number of degrees of freedom is sizeable. On the basis of this simulation, the branch cut contribution can indeed be neglected in the hadron resonance gas approximation to the thermodynamics of a quark–gluon plasma. This weak contribution might be explained by multiple resonances which interfere so strongly that their net sum remains low. A sum of many complex terms must in the end combine to yield a real answer. Our last example in this section hopefully provides a warning about how misleading it can be to represent a spectral function as a sum of infinitely many significantly overlapping peaks.

In the above, we have presented a one-field representation of multiple peak spectral functions. In general, the effective Lagrangian in the background is nonlocal. We have shown that consistent definitions for the number of effective degrees of freedom and for the thermodynamic quantities involved in the energy–momentum tensor are possible and give physically interpretable answers in several cases. The number of degrees of freedom defined by N_{00} is not the number of field components used, which is one, but for narrow widths, the number of peaks in the spectral function. In particular, for infinite lifetimes (zero widths), the resonance gas approximation would be exact. For very short lifetimes (infinite widths), this quantity approaches 1, the number of field components: everything melts into a single physical object.

References

1. T.S. Biró, S.G. Matinyan, B. Müller, *Chaos in Gauge Field Theory* (World Scientific, Singapore, 1985)
2. S.G. Matinyan, G.K. Savvidy, N.G. Ter-Arutyunyuan-Savvidy, Classical Yang-Mills mechanics, nonlinear color oscillations. Sov. Phys. JETP **80**, 830 (1981)
3. S.G. Matinyan, B. Müller, Quantum fluctuations and dynamical chaos: an effective potential approach. Found. Phys. **27**, 1237 (1997)
4. T.S. Biró, S.G. Matinyan, B. Müller, Chaos driven by soft-hard mode coupling in thermal Yang-Mills theory. Phys. Lett. B **362**, 29 (1995)
5. CMS Collaboration, Charged particle multiplicity. JHEP **08**, 141 (2011)
6. CMS Collaboration, Elliptic flow and low-pt spectra. Phys. Rev. C **87**, 014902 (2013)
7. CMS Collaboration, Nuclear effects in transverse momentum spectra of charged particles in pPb collisions at $\sqrt{s_{NN}} = 5.02$ TeV. EPJ C **75**, 237 (2015)
8. ATLAS Collaboration, Measurement of charged-particle spectra in Pb+Pb and pPb collisons at $\sqrt{s_{NN}} = 2.76$ TeV with the ATLAS detector at the LHC. JHEP **09**, 050 (2015)
9. ALICE Collaboration, Pseudorapidity and transverse-momentum distributions of charged particles in proton–proton collisions at $\sqrt{s} = 13$ TeV. Phys. Lett. B **753**, 319 (2016)
10. ALICE Collaboration, Energy dependence of the transverse momentum distribution of charged particles in pp collisions measured by ALICE. APJ C **73**, 2662 (2013)
11. R. Hagedorn, Nuovo Cimento Suppl. **3**, 147 (1965)
12. V. Pareto, La courbe de la répartition de la richesse (Orig. pub. 1896), in *Œuvres complètes de Vilfredo Pareto*, ed. by G. Busino (Librairie Droz, Geneva, 1965)
13. R. Koch, *Living the 80/20 Way: Work Less, Worry Less, Succeed More, Enjoy More* (Nicholas Bearley Publishing, London, 2004)
14. W.J. Reed, The Pareto, Zipf and other power laws. Econ. Lett. **74**, 15 (2001)
15. C. Tsallis, *Introduction to Nonextensive Statistical Mechanics* (Springer, Berlin, 2009)
16. C. Tsallis, Nonadditive entropy: the concept and its use. EPJ A **40**, 257 (2009)
17. C.-Y. Wong, G. Wilk, L.J.L. Cirto, C. Tsallis, From QCD-based hard-scattering to nonextensive statistical mechanical description of transverse momentum spectra in high-energy pp and $p\overline{p}$ collisions. Phys. Rev. D **91**, 114027 (2015)
18. M. Biyajima, T. Mizoguchi, N. Nakajima, N. Suzuki, G. Wilk, Modified Hagedorn formula including temperature fluctuation – estimation of temperatures at RHIC experiments. EPJ C **48**, 597 (2006)
19. T.S. Biro, Z. Schram, L. Jenkovszky, Entropy production during hadronization of a quark-gluon plasma. EPJ A **54**, 17 (2018)
20. M.M. Homor, A. Jakovac, Particle yields from numerical simulations. Phys. Rev. D **97**, 074504 (2018)
21. M. Gyulassy, M. Plümer, Jet quenching in dense matter. Phys. Lett. B **243**, 432 (1990)
22. X.N. Wang, M. Gyulassy, Gluon shadowing and jet quenching in A+A collisions at $\sqrt{s} = 200$ GeV. Phys. Rev. Lett. **68**, 1480 (1992)
23. X.N. Wang, Z. Huang, I. Sarcevic, Jet quenching in opposite direction of a tagged photon in high-energy heavy-ion collisions. Phys. Rev. Lett. **77**, 231 (1996)
24. Z.B. Kang, I. Vitev, H. Xing, Vector-boson-tagged jet production in heavy ion collisions at energies available at the CERN Large Hadron Collider. Phys. Rev. C **96**, 014912 (2017)
25. B. Müller, *The Physics of the Quark Gluon Plasma*. Springer Lecture Notes, vol. 225 (Springer, Berlin, 1985)
26. C.E. DeTar, J.F. Donoghue, Bag models of hadrons. Ann. Rev. Nucl. Part. Sci. **33**, 235 (1983)
27. Y. Aoki, Z. Fodor, S.D. Katz, K.K. Szabo, The equation of state in lattice QCD: with physical quark masses towards the continuum limit. JHEP **2006**(01), 089 (2006)
28. M. Creutz, Phase diagrams for coupled spin-gauge systems. Phys. Rev. D **21**, 1006 (1980)
29. H.J. Rothe, *Lattice Gauge Theories*. World Scientific Lecture Notes in Physics, vol. 82, 4th edn. (2012)

30. A. Jakovac, A. Patkos, *Resummation and Renormalization in Effective Theories of Particle Physics*. Lecture Notes in Physics, vol. 912 (Springer, Berlin, 2016)
31. A. Jakovac, Representation of spectral functions and thermodynamics. Phys. Rev. D **86**, 085007 (2012)
32. A. Jakovac, Hadron melting and QCD thermodynamics. Phys. Rev. D **88**, 065012 (2013)
33. T.S. Biro, A. Jakovac, QCD above T_c: hadrons, partons and the continuum. Phys. Rev. D **90**, 094029 (2014)
34. T.S. Biro, A. Jakovac, Z. Schram, Nuclear and quark matter at high temperature. EPJ A **53**, 52 (2017)
35. F. Karsch, K. Redlich, A. Tawfik, Thermodynamics at nonzero baryon number density: a comparison of lattice and hadron resonance gas model. Phys. Lett. B **571**, 67 (2003)
36. A. Tawfik, QCD phase diagram: a comparison of lattice and hadron resonance gas model. Phys. Rev. D **71**, 054502 (2005)
37. V.V. Begun, M.I. Gorenstein, M. Hauer, V.P. Konchakovski, O.S. Zozulya, Multiplicity fluctuations in hadron-resonance gas. Phys. Rev. C **74**, 044903 (2006)
38. T. Steinert, W. Cassing, Covariant interacting hadron-resonance gas model. Phys. Rev. C **98**, 014908 (2018)
39. P. Huovinen, P. Petreczky, Hadron resonance gas with repulsive interactions. J. Phys. Conf. Ser. **1070**, 012004 (2018)

Closing Remarks

We have outlined in this book arguments in favor of transgressing the Boltzmannian density matrix paradigm in quantum field theory, due to an elementary emergence of complexity. We have discussed phase space restrictions and colored noise (dependent on subsystem energy) as examples replacing the $e^{-\beta H}$ factor with something else. In order to provide a background to the mathematical treatment of these and related phenomena we have described the Keldysh formalism and presented examples of its use, emphasizing in particular the role of non-equilibrium spectral functions.

Naturally, a number of problems which would be worth investigating have not been touched upon. Just to mention a few, we have not given examples of open quantum systems or complex network dynamics. For the latter, a field theory with nonlocal action would appear to be necessary. Nothing is known about the renormalization procedure that should be applied in those theories. Furthermore, we have not yet formulated a full treatment of quantum field theory with general relativistic Lagrangian; so far we have only considered a fixed spacetime background, curved but without back-reaction from the quantum fields. But can we understand the emergence of the Unruh temperature, and in particular its appearance as a Hawking temperature at event horizons surrounding black holes, without taking this step? Can we get an intellectual grasp of the quantum uncertainty temperature and a canonical distribution parameter for quark and gluon partonic distribution inside a highly accelerated proton without that?

We have to close this brief discussion without seeking answers to such questions, under the assumption that forthcoming generations of keen researchers will attempt to answer them in the not too distant future.

T. S. Biró and A. Jakovác, *Emergence of Temperature in Examples and Related Nuisances in Field Theory*, SpringerBriefs in Physics, https://doi.org/10.1007/978-3-030-11689-7

Printed in the United States
By Bookmasters